ALSO BY LARRY KILHAM

Non-Fiction
Great Idea to a Great Company: Making Inventions Pay
Winter of the Genomes

Fiction, based on AI
Love Byte
A Viral Affair: Surviving the Pandemic
Saving Juno

MegaMinds

Creativity and Invention

Larry Kilham

Cover photo: art_of_sun/Shutterstock.com

Published by
FutureBooks.info
Available for purchase from:
CreateSpace.com
Amazon.com
and other retailers

ISBN: 1505920957
ISBN 13: 9781505920956
Library of Congress Control Number: 2015901030
CreateSpace Independent Publishing Platform
North Charleston, South Carolina

To my father, a brilliant inventor and product designer, who was my inspiration to write this book.

CONTENTS

Part 2
Big Data and New Solutions

INTRODUCTION

The year is 2050. The oceans have risen, populations are panicking, the simmering climate changes the tropics to deserts, and key resources are nearly depleted. Major governments are calling for a concerted rescue effort whose magnitude will dwarf World War II. Bring in a genius and do something! If we could resurrect Leonardo da Vinci, what would he do?

Haven't we all wished for instant genius when confronted with an apparently intractable problem? Couldn't a genie be released from a bottle and poof! Problem solved. Acts of genius, however, call for more than a genie or even a Leonardo da Vinci acting alone. The ancient Greeks would enlist the aid of sympathetic gods and muses. The Pentagon would tap into its anonymous think tanks where geniuses are thought to reside. The geeks would swarm to the mega computers whose power is predicted to exceed at least a billion humans' combined mental capabilities.

Sometimes the unexpected provides a reprieve. On Valentine's Day, February 14, 2014, Indonesia's Mount Kelud Volcano erupted, killing three and forcing over 100,000 people to evacuate. The belching volcano injected sulfur dioxide seventeen miles into the stratosphere. The condensate from the reaction with water produced sulfate aerosols from sulfuric acid. These tiny droplets reflected incoming sunlight to help cool the earth a little. This was the latest in a series of volcanic eruptions from 2000 onward, which may well have slowed the pace of global warming.

Observing the Mount Kelud phenomenon led to geoengineering proposals to create a stratospheric layer of sulfate aerosols to cover the earth to halt global warming. Starting the process is relatively easy. Small amounts of easily synthesized chemicals would be required, and they could be injected into the stratosphere by conventional military aircraft. I think Leonardo da Vinci would be intrigued by this solution.

Then the problems begin, including life-supporting monsoons greatly reduced in Asia and Africa, further erosion of the ozone layer, and increased catastrophic ocean acidification from the addition of acid rain. The atmospheric chemistry could change in many ways.

As intriguing as this proposed solution may sound, as a practical matter, the solution to the global warming problem will require many solutions. Some will be political, some will be social, some economic, and there will have to be many contributions from creative inventors and engineers. They will have to chip away at challenges ranging from low-cost and pollution-free energy sources to reliable and safe replication of the volcano aerosol cooling effect. Ameliorating global warming is but one challenge for creative and inventive problem solvers.

This book is about the creativity and invention, and how the creative process is enhanced by the availability and accessibility of information in the digital age. It will help you achieve your creative maximum using the entire Internet and other data resources available. Its observations apply to lone individuals, research teams, and everyone in-between.

This book should also help you think about the education of today's kids and all generations to follow. The Web can be an

opiate, trash pile, candy store or inexhaustible resource. Much
of education will be teaching the new generations how to use it
and how to question its results.

We will explore imagination and creativity in this emerging
era of Web-based collective intelligence. The resource rich Web
can help us imagine and create in new ways, and it facilitates
the inclusion of more information and the connected intelligence of other people. Computers now have limitless memory
capacity, plain language instant communication, and software
that parallel the natural thinking process. Computer-based
systems including the Web are powerful assistants for thinking beyond each person's own educational and environmental
constraints.

While we are trying to formulate theories about creativity,
imagination, and consciousness, scientists are rapidly increasing insights and knowledge about how the brain works. This
is one of the last frontiers of earthly science, and the findings
are arriving just in time. Most importantly for this book, we are
learning how the brain predicts and imagines. This branch of
cognitive science is in its early stages, but progress is encouraging. Key to understanding how the brain thinks and acts
are disciplines ranging from neural genetics to functional
magnetic resonance imaging, a kind of brain scanning often
referred to as fMRI.

Now that many people are using massive Web-based data
programs like Google for their research and innovation, will
there be more connected intelligence and collectivization of
information? Has evolution of the imagining, creative mind
slowed down and is it being replaced or greatly augmented
by plugging the brain into the computer clouds? Or are we in
the throes of a paradigm shift not unlike the enlightenment
following the introduction of the printing press? Will humans
remain in control of the creative process?

XIII

Thinking about thinking has always fascinated me. It started when I was in graduate school at MIT in the mid-'60s where I did some interesting and original research on artificial intelligence. This focused on associative inference as a logical way of arranging information in forms that can improve intelligence when new information is gathered. Associative inference is central to some current theories about human memory organization and creativity.

One day, at a MIT campus gathering, I was asked whether I saw the future preferred mode to be human or computer computation and problem solving. I answered, "The computer should do what it does best and the same for the human." It seemed to me like a patronizing and simplistic statement, but to my amazement, the group thought it was rather profound. I think this answer still is valid today, and its recognition is often lost in current debates about the coming impact of AI in society.

The major sea change in recent decades is the pervasive use of the Internet. There is a growing awareness that the significant problems of our time are so complex that creative solutions by lone geniuses are still possible but less and less likely. Search engines have brought us worldwide databases searchable from almost any direction of inquiry. Information is the commodity, and its direction and use is still under control of man.

There is a pressing challenge to deal with computer and information issues now because we have been in materials and energy ages and now these resources are disappearing. Much of the technology advantage to be exploited is through mega thinking and intellectual breakthroughs. How do we promote powerful new thinking? What can be done to head off the many basic problems confronting civilization? Indeed, how do ordinary people avoid lives of being couch potatoes and instead seek joy and productivity through exploration of the unknown and through original thinking?

I have come to the conclusion, to be explored and explained in this book, that at least in theory a thinking machine can be devised for any thinking task that can be done by a human, but the stupendous scale of everything required limits its application in many challenging applications. The optimum solution is to combine humans with the computer cloud working together as connected intelligence. The lone geniuses are as important as ever but increasingly in a role as the thought process manager imbedded in a sea of data and computing power.

I will also explore what I perceive to be the key attributes of the creative mind. This takes us to a space apparently off-limits to computers: projecting the mind out of most all of its frame of references and rethinking a problem in the frame of reference of the problem itself. The mind is projected to a frame of reference most suitable for visualizing the invention or creation. This requires a bright mind, but also essential are years of practice in conditioning the mind for this mode of thinking and aggressively gathering information for areas of continuing interest.

Many of my insights about the creative mind come from my own experiences. I come from an extended family of artists, inventors and entrepreneurs, and I will draw heavily from what I have learned in my own projects and from my family. I will also provide a solid research base for my ideas through citations from recent research on cognitive processes, computation and related areas.

This book progresses through three areas of revelation about the creative mind. It reviews the history of invention and societal condition encouraging invention. There is a review of the genius mind and creativity. This looks at artists, composers, physicists and inventors. Then we examine the extraordinarily creative mind—think Einstein—which I call the projected mind.

MegaMinds goes on to discuss experiments with computers to think like humans in a very high tech field called artificial intelligence (AI). We then look at recent research findings about the human brain and how it absorbs information, thinks and creates. We will look for parallels between AI and the human thought processes. We will explore creativity and invention using Google.

In its last chapters, the book then moves on to the emerging science of very large databases. When we have millions of bits of data, we can move into new methods of analysis to find patterns in the data sets. It will sort of look like the human brain all over again, but done in the computer clouds. Google and others have found that there are new and more efficient ways to check spelling, translate languages, discover drugs, design aircraft and do many other things. We will examine the use of the computer clouds for analyzing analysis of data and formulating of conclusions through the emerging field of collective intelligence.

The concluding chapter looks ahead with a vision of about how the Internet and digital age will affect humanity for years to come. I do not foresee computers and AI taking over the world, but I do see major areas of caution.

Fasten your seat belts!

PART 1

THE ESSENCE OF CREATIVITY AND INVENTION

NEW CREATIVITY FOR RENEWED PROSPERITY

If the presence of electricity can be made visible in any part of the circuit, I see no reason why intelligence may not be transmitted instantaneously by electricity.

Samuel F. B. Morse, American artist and inventor

On a recent flight to Orlando, home of Disney World and fantasy, I sat next to a bright young entrepreneur. Working in the automobile parts supply business, he learned who the buyers are and who the sellers are, pricing, key trade shows to meet important contacts and so forth. He coupled this knowledge and his boundless inquisitiveness and energy with the new marketing medium of the Internet to start an "after-market" automobile accessories supplier mainly to the world of Toyota.

He has about 20 employees, is self-financed with no debt or government aid, has an adoring wife and great kids, and is generally enjoying the American Dream. He is less than half my age and, I thought, an ideal observer of the enterprise world coming upon us.

I asked him, "Are you developing any of your own products? Will you have some cool addition to the standard stuff everybody is making and selling?" He thought for a minute and replied, "There are extremely few, if any, new inventions anymore."

You might think we will turn into the Internet marketing capital of the world and leave it to the rising class of Asian engineers who have found stimulating opportunities in their home countries to generate the products the whole world will need for today and tomorrow. Creativity will be reduced to thinking up new Corporate Goals statements and new luncheon theme ideas.

But Bette Nesmith Graham didn't know this. A single mother and secretary in Dallas, she thought there would be a better way to cover up mistakes made in typing. She recalled her long ago artistic experience and looked for a liquid mixture to paint over the typing errors. She made the first formulations in her kitchen blender.

In 1956, Ms. Graham founded the Mistake Out Company, later well known as Liquid Paper or "white out," starting on the proverbial shoestring and working nights and weekends. By 1968, she had her own plant and 19 employees. She sold her company for $47.5 million. This is still possible at all levels from the kitchen chemistry lab to the killer app corporate development project or to the multinational research initiative. In 1993, I started my last company on the dining room table soldering parts together mostly purchased at the local Radio Shack. It grew into a small but leading gas detection instrument company, which I sold in 2007 at many times my investment.

We are in new times and uncharted territory in the saga of enterprise. The United States and the rest of the western world are facing the possibility of no growth or at best very controlled growth for decades or longer. Major product ideas and resources are harder to find. We must create new solutions, products and services as a major component of future sales.

Entrepreneurs and managers must rediscover discovery. For the last few decades, the mantra in business schools and corporate training programs has been marketing. This has been the way to the top. Information technology has also become an important fast track for rising managers. In many cases, product development and intellectual property accumulation has been a discretionary activity, seriously pursued when extra cash was in great supply.

This book focuses on the mindset and creative process involved to imagine, create and invent in the twenty-first century. This subject is not generally taught in schools and colleges, and it does not lend itself to a few simple rules for success, but we must tackle it if we are going to enjoy renewed prosperity any time soon.

While management of innovation has been a popular management development subject, the creative process itself is often not meaningfully addressed. This book will help you identify creative innovators, help you know what is a good creative environment, and help you understand what knowledge resources innovators need to carry out the creative process. This information is equally valuable for the self-guided creator and entrepreneur. I will combine observations of historically creative and inventive people, new findings from cognitive science about the creative process, and ways to use the Internet and computer clouds to greatly enhance success in a creative project.

Whatever the era or product, the successful project or company starts with a creative visionary—somebody who is imaginative and persistent and who has a multifaceted mind.

Would an American corporation in the early 1800's (or now) hire as their chief designer a financially failing artist with radical political views and an itchy foot for world travel? There was such a person. He did not have a comfortable job, but he had a vision to develop a communication system that could send messages faster than the best steam trains and ships and unhindered by rain, sleet or snow. He was Samuel Finley Breese Morse, a distant relative, who invented the telegraph.

The communications revolution, still in progress today, started with Morse's telegraph key.

In 1832, while on a sea voyage back to America, Morse began to think about the concept of a telegraph system. He knew the basic principles of electro-magnetism, but not the practical aspect of engineering products and systems. Several European inventors were also working on telegraph systems, but apparently their efforts were unknown to him.

Morse used his creative abilities to see relationships and possibilities. His breakthrough was coding letters and numbers as groupings of binary digits. This allowed the simplicity of sending messages over one wire (the return circuit being ground) instead of several wires that would be required for simple or no coding schemes. The competing European designs required as many as 35 wires.

Public demonstrations of telegraphy happened about 12 years after Morse's first vision of it. What carried him through those wrenching times was perseverance, the ability to tinker and improvise, a wide scope of thinking about all aspects of the design, and his ability to bring other people to help when design, manufacturing and other challenges required additional talents and facilities.

Both Bette Nesmith Graham and Samuel F. B. Morse were iconic American inventors who illustrate traits in common that will be valuable to anyone interested in creating new designs and products:

- **Unleash your curiosity, quest for knowledge, and propensity for noticing things.** No lesser minds than Leonardo da Vinci and Albert Einstein were noted for being passionately curious, using their imagination as their prime lens to see ahead and their creativity to solve problems. Einstein wrote, "The important thing is not to stop questioning." You should also notice all kinds of things, however unrelated to your quest they may seem. When Willis Carrier noticed the apparently odd behavior of water droplets in fog, he had stumbled into the basics of the novel technology of the Carrier Corporation, world leader in air conditioning.

- **Project your mind into imagination space, focusing on all the interrelated aspects of what you are creating or inventing.** To create your Eureka! moment, you must forcefully move your mind beyond the existing thinking about the subject. You must move out of your conscious world and focus your mind in a new place occupied only by the new creation. This is your glorious imagination space. Some people, very few, keep this imaginative ability through adulthood. Their imaginings lead to inventions, art, designs and explorations of many frontiers never seen before. To start, try to be a child with the almost naïve capability of unfettered imagination. Emotion is part of this creative formula, and it has not been replicated in any advanced computer.

- **Bring in experts and specialists whenever and wherever appropriate.** A common mistake is to be overly protective about your novel idea. At the earliest possible time, you should have your design or composition reviewed by an associate, faculty member, consultant or other trustworthy knowledgeable advisor. Usually you do not have to disclose important details to protect from copying, and very often, a reviewer can give you surprisingly good guidance on design or composition improvement.

- **Focus on the practical, useful, needed and beautiful.** Very often, inventions and other creations start out answering to a major need or a broad interest. Then the project morphs into a personal passion with little or no market value. Whether you are a garage tinkerer or Thomas Edison, ultimately your commercial success depends on developing something which economically fills a real need and which looks attractive to potential buyers. As you develop prototypes, theories or compositions, show them to people in the market for feedback about your invention's overall attractiveness.

- **Be persistent. Do not give up!** Thomas Edison had not developed a good prototype light bulb even after 9,000 experiments. For major inventions like the light bulb, there is more lonely persistence than glamour. Even minor inventions seem to take more time than imagined to get to the demonstration prototype stage.

These key parts of creative thinking will be explored in part one of this book.

Fast forward to 2005. Steve Jobs, the legendary leader at Apple, is initiating a great leap forward. He has directed about 200 of his best engineers to create what we now know as the iPhone. Like Morse, he is not the first with some version of his product. And like Morse, Jobs can focus on a product vision that combines needs satisfaction, functionality, apparent simplicity, and, in addition, design beauty. The resultant product is a combination of invention, engineering, and aesthetic appeal. In short, it is a bold act of creativity.

Where the telegraph initiated the era of wired communications, the iPhone has put the computer clouds (almost infinitely large bundles of data and services available by Internet) in the palm of your hand. The telephone is not obsolete, music radio will not go away, computers of all sizes will always be here, video games will always be ready to use, and data transmission will always be available through specialty equipment; but now all of these modalities are available together through a personal portable device.

**Steve Jobs proudly showing
his new MacBook Air creation**

The iPhones and similar smart phones are forever changing the way we use computers and communications. Now there are many competitors to the iPhone, but the design led by Jobs crystallized that this new communications and computing package was not a temporary success. It is a basic paradigm shift with benefits for everyone.

In retrospect, the design requirements seem obvious enough:

- Use a powerful operating system that does not hog memory (easier said than done).
- Develop reliable touch screen controls instead of typing keys (also no design job for amateurs).
- Enchant the prospective buyer with a beautiful design with no keys, shiny aluminum instead of black plastic, and colorful icons.
- Offer a huge music and apps library.

Apple's sales of the iPhone have skyrocketed from nothing in early 2007 to 17% of the world market in late 2009. Steve Jobs commented in November 2009, "We're making our most innovative products ever, and our customers are responding. We're thrilled to have sold of 5.2 million iPhones during the quarter, and users have downloaded more than 1.5 billion applications from our App Store in its first year." In 2013, two years after Steve Jobs' death, Apple shipped 153 million iPhones and accounted for 15% of one billion total market shipments. Four other companies accounted for most of the rest of the smartphone sales.

Samuel F. B. Morse of course did not have the technology and resources available to Jobs for his design project. Most important for Apple is the role of computers in complex design. The several hundred engineers assigned to the project could not integrate all the subsystems of the iPhone such as the radio circuits, internal power supply, microprocessor, software, touch screen display and mechanical packaging without computerized integration of the subsystem designs. The search for components and design solutions would require intense use of the Internet.

Still, even in the Age of Google, a visionary leader is required, and Steve Jobs is reported to have mercilessly driven his design group, never taking "no" for an answer. There were screaming matches in the hallways, doors slamming and completely burned out engineers.

The iPhone and many other recent developments from tiny pills to giant airliners call for the new tools for creativity, invention and design which we will explore in parts 1 and 2 of this book. A common thread among these tools is computer clouds and computer networks. They show up as:

- Use of large research teams in virtual labs defined by computer networks.
- Artificial intelligence (AI).
- Novel methods of analysis of massive data sets or "big data."
- Collective intelligence involving communities larger than research teams sharing private computer network wikis.
- Designing and inventing with the help of Google and other search engines.

These creative tools are required because the most creative challenges are much more complex compared to a century ago. The development of the iPhone required thinking in a much more complex space than did the development of telegraphy even though both were hugely important in their time.

I will illustrate my story by references to thinkers and creators ranging from Leonardo da Vinci to Thomas Edison. We will see what Mozart and Einstein had in common. We will see what ordinary people can do to enjoy a more creative and prosperous life. There are many examples ranging from Bette Nesmith Graham to various enterprises in my own family to draw from.

We will go back to basics by reinventing the wheel; see how to design the world's most successful bird feeder; discover that computers can design electronic circuits; see how Pfizer's scientists collaborate on online shared research ideas; examine Boeing's new design approach for the 787 Dreamliner; learn that an oil spill challenge was solved by a concrete expert found by an inventor's outreach on the Internet; and see how MIT is approaching the climate change analysis problem by using an online "collaboratorium" of collective intelligence from many researchers in hundreds of fields.

I hope that somewhere in this stream of stories you will find something to enhance your creative life—maybe even make you insanely rich. Maybe there will be ideas here to help your children launch creative and remunerative careers. At the organization level, the ideas in this book could benefit employees in order to achieve greater profitability for the company or productivity for the organization.

This book is about the creative process, and how creativity and invention is enhanced by the availability and accessibility of information in the Internet age. It will help you achieve your creative maximum using all of the Internet and other data resources available. Its observations apply to lone individuals, teams large enough to put a man on the moon, and everything in-between.

We will explore creativity and invention in this emerging era of Web-based collective intelligence with its almost infinite ability to connect to others and help us imagine and create in new ways. With limitless memory capacity, plain language instant communication, and software that parallels the natural thinking process, computer-based systems including the Web, enable humanity to transcend each person's own educational and environmental constraints.

Recent years have seen an explosion in new understanding about the human brain. New reports are appearing weekly in scientific and general media. Consequently, how the brain predicts and imagines is becoming clearer. At the same time, technologies that are more sophisticated are being developed to approximate human thought. Quite likely, any discoveries will have ramifications in cyberspace and for our creativity and invention strategies.

More pervasive still—at least right now—are the almost infinitely large computer clouds with libraries of trillions of pieces of data all of which are accessible by the click of a mouse. The clouds are everywhere. They are behind everything from online weather forecasts to harboring Facebook, Amazon, and Google. Advanced use of the computer clouds is for collective intelligence and data search for online inventing. The clouds support the research and creativity required for solving some of our greatest mysteries and problems, all of which seemed to have been too complex to deal with. This book will review these developments and help you develop your techniques for online creativity.

2

LEARNING FROM THE MASTERS

People make a mistake who think my art has come easily to me. Nobody has devoted so much time and thought to composition as I. There is not a famous master whose music I have not studied over and over.
Wolfgang Amadeus Mozart

I keep the subject of my inquiry constantly before me, and wait till the first dawning opens gradually, little by little, into a full and clear light.
Isaac Newton

My father started instructing me in the dos and don'ts of creative design as soon as I could toddle into his workshop. While the other kids I knew thought I was a little weird, I stuck with the creative path because my father said it would always be remunerative and personally satisfying with a big emphasis on the latter. By the time I was in the world of engineering, creative design was second nature to me, and I could not understand why it did not come as naturally to others.

We lived in the country and when I was six or so, I often passed the lazy summer days fishing. I noticed that the standard red and white striped lures the anglers used did not work as well as they should on a major local fish, pick-erel. These lures were called spoons because they were metal and spoon-shaped minus the handle. The fish seemed to be intrigued with flashing gold things, so I experimentally pro-duced gold spoons by hammering oval pieces of sheet brass on an anvil. I punched holes at both ends for the fishing line and the hook. There were a number of little produc-tion techniques to perfect, which was easy to do because my father had a machine shop on the property, and I wound up selling the lures for the equivalent of several dollars each in today's currency.

From that experience, I learned that you are never too young to discover needs and to invent profitable solutions for them. It can lead to an interesting and remunerative entrepre-neurial career.

In my early teens, I became interested in electronics by making "crystal radios." Using a coil wound on a toilet paper tube for tuning to a station, a safety pin called the cat whisker, a chunk of galena crystal ordered through the mail, head-phones and a very long antenna you could listen to the local radio station (for a good description and illustrations of these early radios see Crystal Radio on Wikipedia). The cat whisker and galena, a semiconductor, together were a rudimentary diode. By rectifying the radio frequency signal (producing a DC component of the signal), they could produce a faint audio signal.

Later, I found that a Gillette razor blade worked just as well as the purchased galena. In fact, I serendipitously discovered that with a second cat whisker and a battery you could amplify the radio signal. I was too young and naïve to understand that I may have discovered the transistor before it was invented!

This experimentation led to an interest in "ham radio" where with receivers, transmitters, and an official license, you could telegraph by Morse code or even talk by microphone all over the world. Eye-popping stuff for a country boy. I eventually designed my own circuits, drawing upon intuitive modifications of circuits found in ham radio magazines. Years later, I would learn how to design circuits using electrical engineering theory, but that was never as satisfying. It all paid off. My constant development of practical circuit designs led to the products for several instrumentation companies that I started and sold.

If children are to be encouraged to pursue inventive or entrepreneurial careers, there should be an environment encouraging intellectual inquiry and creative thinking at an early age. Usually a parent, teacher or fellow student is encouraging and mentoring the budding genius. This process often starts before six years of age. The genius-in-the-making must constantly be searching for knowledge, improvement and opportunities for outstanding intellectual achievements.

For major achievement by creative people, there is continuous and demanding intellectual work involved for most of a lifetime. This usually works best for self-starters and those who can maintain intense focus on their projects.

You or any adult can be a creator and inventor starting at any point in your life. For success, however, you must stick with your creative and inventive skills. You must acknowledge that the inevitable downturns are opportunities that come along to learn something valuable for future success.

It has been no different for history's well-known geniuses. They did not start composing great symphonies or invent light bulbs by short bursts of creativity. The initial insight may have been similar to noticing something about the apple falling from the tree, but the execution to even a prototype production required lots of research and perfection—possibly years of ups and downs.

Wolfgang Mozart and Pablo Casals

Both Mozart and Casals had fathers who introduced them to their art long before most other children their age would be thinking seriously about getting involved in music. This syndrome appears with many other great artists, engineers, inventors, and scientists. The father or mother who is comfortable and accomplished in a profession encourages their son or daughter to be a great achiever in that field—hopefully to be the best.

Wolfgang Amadeus Mozart (1756-1791) was born into a musical family in Salzburg, Austria, and at five years old, he could read and write music. Wolfgang's father, Leopold, was a composer and teacher, and he wrote a violin textbook. When Wolfgang was four his father began teaching him music. He eventually home-schooled Mozart and his sister Nannerl. Wolfgang Mozart had a near photographic memory, and he learned a lot by mimicry. A famous example is his hearing Gregorio Allegri's *Miserere* once in the Sistine Chapel and later writing it down entirely from memory.

What was Mozart's genius contribution? According to the *Music Encyclopedia:*

To the neat and symmetrical periods of the Haydn symphony and sonata, with their fresh thematic treatment, Mozart added a tender grace and sweetness like the conceptions of a Raphael in painting. He was the apostle of melody. If he had never written, the art of music would have remained something quite different from what we know it. And wherever there are lovers of refined, noble melody, there will the music of Mozart be loved.

Mozart was not all sweetness and tenderness, however. He was given to screaming shocking profanity in any company; he was paranoid; and he was given to twitching, tapping, jumping around, and making faces. Was this just artistic energy? It is generally believed he had attention deficit hyperactivity disorder (ADHD), which is an affliction of many intensely creative people. He also had, in addition to everything else, obsessive-compulsive disorder. I will explore mental abnormalities in creative people in chapter four.

Cellist Pablo Casals (1876-1973) also started his immersion in music very early. His father was an organist and choirmaster who instructed Casals in various instruments. At four, Casals could play the violin, flute and piano, and at eleven, he started his lifelong specialty of the Cello. A key trait of the genius mentality is illustrated when asked at the age of 94 why he still practiced every day, he answered, "Because I think I detect signs of improvement."

Leonardo da Vinci

If the parents are not prepared to give the child a great intellectual or career start at a very early age, then he or she may still be given a fast start at a suitable school. Such a school might well be very vocationally oriented. This was the case for Leonardo da Vinci who is considered one of the greatest painters of all time and a brilliant scientist, engineer, and inventor.

Born near Florence, Italy in 1452, Leonardo did not have a promising start, being the illegitimate son of a notary and a peasant woman. (I generally use "Leonardo" rather than "da Vinci" when writing about Leonardo da Vinci, following the practice of most of the writers about him.) This social stigma prevented him from enrolling in an advanced classical academy, but at the age of fourteen, he was admitted to the workshop of Andrea del Verrocchio, probably the best place in Florence to be an apprentice under a craftsman. His talented fellow students, including Botticelli, Perugino, and Lorenzo di Credi, would also have influenced him.

Leonardo's father was encouraging and helpful in getting him this far. In 1472, at the age of 20, Leonardo was registered in the Florentine painters *Guild of St. Luke.* Shortly after that, he completed his first solo painting, *The Annunciation.* Many more masterpieces followed including *The Last Supper* and *Mona Lisa.*

Leonardo's intellectual attributes included those in common with most other very creative people: intense curiosity, an insatiable quest for knowledge, a large memory, intense concentration, and continuous improvement of his knowledge and skills. He was always looking for ways to relate diverse categories of information and to search for novel patterns and insights.

Leonardo pursued parallel creative efforts in drawing, anatomy, science, optics, machinery, and engineering. Leonardo saw no problem doing both "art" and "science" because to him these were complementary and mutually supportive skills. Art was the skill in doing something or in mastering a field of study, such as the art of medicine; and science was the knowledge of living forms, their intrinsic nature, and their underlying principles.

Leonardo's imagination was tied to an understanding of nature. This gave him an aesthetic feeling of beauty and natural forms of design. This, in turn, gave him a reassuring feeling of correctness in his synthesis and analysis. In addition, Leonardo thought in terms of *fantasia,* the artist's creative imagination.

Whether Leonardo would relate so much of his thinking to nature if he were living today, is an open question. Would he seize upon computers, the Internet, jet travel, and artificial environments as useful conveniences and aids to thinking, or would he think of these appurtenances mostly as needless destructors of nature?

I think Leonardo would embrace these modern technologies because he certainly made major efforts to improve life in his time through technology and engineering. For example, he designed artificial harbors and thought about the rerouting of rivers. In addition, he is famous for his proposed designs of flying machines.

I also think Leonardo would agree, were he alive today, that with our Web incorporating billions of websites and trillions of pieces of information, with access to all by a laptop computer, that this data and computing cloud is closer to his hope for access to a nearly infinite universe than simply being able to see what can be observed from nature. Leonardo, however, would not give up observing nature. Everything would count.

While scientific instruments to aid in observing nature were available in the Renaissance, they did not see down to the molecular level nor out to the galaxies. Routine observation of things and phenomena beyond direct human experience did not happen until the beginning of the twentieth century. These capabilities led to new basic branches of science including quantum mechanics, microbiology and astrophysics.

There are thousands of specialty branches of science and together they interconnect to form a universe whose size and scope are beyond even what Leonardo could comprehend. Fritjof Capra, a leading historian of Leonardo noted:

The Scientific Revolution replaced the Aristotelian worldview with the concept of the world as a machine. From then on the mechanistic approach—the study of matter, quantities, and constituents—dominated Western science. Only in the twentieth century did the limits of Newtonian science become fully apparent, and the mechanistic Cartesian worldview begin to give way to a holistic and ecological view not unlike that developed by Leonardo da Vinci. With the rise of systemic thinking and its emphasis on networks, complexity, and patterns of organization, we can now more fully appreciate the power of Leonardo's science and its relevance for the modern era.

It is true that the current world community dedicated to solving the remaining big problems have redefined the analytical methodologies and philosophies they work with. The remaining challenges are incredibly complex and include how living, replicating molecules evolved out of cold, lifeless chemistries; understanding the mechanisms of consciousness and free will; and creating the ontology of very large thinking computers. Many have characterized this as a movement away from a strictly mechanistic, Cartesian view of the world to a holistic view involving boundless complexity and ecology.

The universe is not seen as neat separable systems to be described by exact mathematical models. Instead, the universe is modeled as many interrelated systems whose behavior is modeled by active elements in those systems (often called "agents") and whose behavior is often described in probabilistic terms. A new science has emerged based on models of complexity and is the focus of the Santa Fe Institute in Santa Fe, NM, USA.

If Leonardo was resurrected and dusted off to help solve the mega problems we are struggling with today, how would he go about it? My guess is that Leonardo would be handed a laptop and told to log into Google to explore the vast data terrain. Leonardo would protest that he learned his skills in a *bottega* workshop where there was a constant exchange of ideas between fellow artisans, mechanics and inventors.

I do not know exactly what Leonardo would do except that because he would relish the challenge, he would pursue a solution with great energy. Would he act alone in classic hero fashion? Form a team of advisory experts? Be very plugged in to the Web? We cannot know exactly how he would react, but we can say that it would not be in his nature to subdue his curiosity nor to refuse to look at any information from any source. He would not seek the easy way out. Leonardo would want his solution to be original, responsive, and truthful.

With the intellectual playing field flattened by giving Leonardo the latest Web tools, can we assume that he would excel or even triumph with singular success in providing a creative solution using the computer clouds? The answer is he would do fine. Leonardo da Vinci and the computer clouds would be a hard combination to beat.

Albert Einstein

As I review the salient aspects of many acknowledged geniuses, the closest match in many ways to Leonardo da Vinci I can find is Albert Einstein. Their research and fields of interest were in different areas, but their points of view about life, science, creativity and other related areas are very similar.

Leonardo said:

> The ambitious, who are not content with the gifts of life and the beauty of the world, are given the penitence of ruining their own lives and never possessing the utility and beauty of the world.

Einstein stated it this way:

> The ideals that have lighted my way, and time after time have given me new courage to face life cheerfully, have been Kindness, Beauty and Truth...The trite objects of human efforts—possessions, outward success, luxury— have always seemed to me contemptible.

Both men's developmental years were similar so we might expect similar outlooks in life. Albert Einstein was born in Ulm, Germany on March 14, 1879. His father was a salesman and engineer of electrical equipment. He probably began Einstein's learning process at home with talk about electro mechanical systems—maybe without Einstein being conscious of this—and this thinking mode would be key to Einstein's visualization of multiple objects moving this way and that in space.

Einstein completed formal schooling in Switzerland with a degree in physics. He chafed at authority and regimen and felt that rote learning reduced creative thought. While he did well in math and physics, and felt that math was the best way to express nature's relationships, he was not considered a brilliant mathematician.

Leonardo de Vinci	Thomas Alva Edison	Albert Einstein
Artist, Scientist, Inventor	Inventor	Scientist
1452-1519	1847-1931	1879-1955

GREAT CREATIVE AND INVENTIVE MINDS
WHAT THEY HAD IN COMMON

Early training and interest in certain skills
Curiosity and imagination
Projecting their mind
Thinking in analogies
Persistence

Both Leonardo and Einstein supported themselves by work relating to their education and interests, Leonardo as a contract artist and engineer, Einstein as a Swiss patent examiner. While both men had moments of engaging, outgoing personality, they both tended to retreat into intensely concentrated thought about the natural world as they saw it or a specific problem they were solving.

Both men avoided collaborative research, needless publishing, or joining collective science groups. Neither would be likely to be a part of sponsored team research prevalent today. Their genius flourished in part because of their intellectual self-sufficiency. The Leonardos and the Einsteins cannot stop their relentless pursuit of knowledge goals, anytime day or night, with or without associates, because they are passionately curious, using their imagination as their prime lens to see ahead and their creativity to solve problems.

Einstein wrote:

> The important thing is not to stop questioning. Curiosity has its own reason for existing. One cannot help but be in awe when he contemplates the mysteries of eternity, of life, of the marvelous structure of reality. It is enough if one tries merely to comprehend a little of this mystery every day. Never lose a holy curiosity.

A professor of mine, Murray Eden, who studied at Princeton University when Einstein was there, tells the following story. One day he was at Einstein's office to pick up an employee and drive her home at the end of the day. Einstein seemed to be reflecting on a major unknown and asked what might be the unknowable: "Why is it that television antennas in America are all horizontal?" This genius, who was a major contributor to electromagnetic theory, had not considered that television broadcasting in America is set up for horizontally polarized radiation, received best by horizontally oriented antennas. (In its beginning, television broadcasting could have been decided to be horizontally or vertically polarized but not both.)

For new inventors the message is: Do not worry about not being an Einstein. If you understand your area of specialty thoroughly, you can be its Einstein.

All descriptions of Einstein talk about his thought experiments. He asked what a light beam would look like if you could ride beside it. This was part of the inspiration of his special theory of relativity. In conceptualizing his general theory of relativity, he imagined riding an elevator accelerating in space. In talking about gravity warping space and time, he describes billiard balls rolling down the surface of a trampoline deformed by a bowling ball.

Another factor in common among the classic geniuses from Leonardo to Einstein is intellectual honesty. First, you must be sure that your work is producing a true answer to what you seek. My father, who was a very prolific inventor with nineteen patents, always counseled me, "Make sure you satisfy yourself first that your creation is true, original, and useful." Information offered on Internet searches offers no guarantees about its intellectual honesty and truth.

Thomas Edison

The 1800-1950 period was a time span of extraordinary inventiveness. There were very smart, dedicated inventors who were lucky to find themselves in an embarrassment of newly emerging resources and know-how in electricity, chemistry, mechanical engineering and more. Theirs was an era of burgeoning economies wresting themselves from agriculture and colonialism as well as developing new equipment for industrialization and wars. Investment money from wealthy individuals seems to have been relatively easy to find. It was a good time to be a practical inventor.

Rather than survey all the brilliant inventors of this period, I will focus on Thomas Alva Edison so that we can see in greater depth how such a person's mind works. Edison was born in 1847 and lived for 75 years. During that time, he amassed 1,093 US patents and additional foreign patents. Edison's inventions were as diverse as communications, the media, electrical machinery and systems, mining technology, and storage batteries.

Edison's only formal education lasted three months, and he described himself as fairly consistently at the foot of his class. Much of his education was home schooling by his mother, and he studied chemistry for a short time at Cooper Union, a college of science and art in New York City. While Edison was never proficient in math, he was a natural experimenter. Isaac Newton, James Watt, inventor of the steam engine, and Sir Humphry Davy, the eminent English chemist, had similar unpromising beginnings.

Edison read constantly in such favorite subjects as astronomy, biology, mechanics, metaphysics, music, physics, electrical science, and political economy. In part to make up for his partial deafness, he became a speed-reader and read books faster than other people skimmed them.

In his diary, however, Edison said:

In trying to perfect a thing, I sometimes run straight up against a granite wall a hundred feet high. If, after trying and trying and trying again, I can't get over it, I turn to something else. Then, someday, it may be months or it may be years later, something is discovered by myself or someone else, or something happens in some part of the world, which I recognize may help me to scale at least part of that wall.

In one famous incident, an associate found Edison at his lab bench surrounded by a sea of experimental storage battery test cells. 9,000 experiments had been carried out with no promising developments. His associate offered condolence, "Isn't it a shame that with the tremendous amount of work you have done, you haven't been able to get any results?" "Results!" Edison replied. "Why, man, I have gotten a lot of results. I know several thousand things that won't work!"

This approach is not outdated nor only for the lab of the home-schooled experimenter. I met a high-level researcher at a major pharmaceutical company who sent thousands of soil sample kits all over the world to see which hitherto unknown compound or microbe returned in this quest would turn out to be the next penicillin. As I write, Northeastern University in Boston reports that its Antimicrobial Discovery Center has discovered a very promising antibiotic in the soil called teixobactin. One bacterium it works against is Staphylococcus aureus, or MRSA, which has become antibiotic resistant and is plaguing hospitals.

With the advent of conducting project research using search engines such as Google, one can comb through millions of Web sites and thousands of data reports of possible interest. Using these search results a trained Edison-like brain can sharpen the search inquiries, analyze the data to search for relevant patterns, or return to other techniques such as laboratory experiments.

Of course, if you ask the right question, you are more likely to get a relevant answer. As F. R. Upton said about Edison: "One of the main impressions left upon me after knowing Mr. Edison for many years is the marvelous accuracy of his guesses."

In his own non-flourishing way, Edison expressed his philosophy of life, and it seems quite similar to Einstein's:

Work. Bringing out the secrets of nature and applying them for the happiness of man. Looking on the brighter side of everything.

As for the value of Edison's genius mind, *The New York Times* of June 24, 1923 reported:

There is one human brain that has a hard cash market value today, in the business and industrial world, of $15,000,000,000. Billions is correct, not millions. That is within 20 per cent of equaling the value of all gold dug from the mines of the earth since America was discovered.

The brain is that of Thomas Alva Edison, who many a time has said to his cronies, "Well, if worst comes to worst, I've got a good trade. I can always make $75 a month as an expert telegraph operator and I can live comfortably on that."

The $15,000,000,000 represents the present investment in America alone in industries which are entirely based on the inventions of Edison or which have been materially stimulated by his inventions. Several of the country's largest industries are included. Incidentally, that figure is $207 billion in 2014 dollars.

Other Characteristics of Creative and Inventive Minds

There are other characteristics of genius inventors. Commonly cited is thinking in metaphors. This is the ability to relate two things that were not apparently related. Such an inventor has the tendency to observe the world with a fresh eye. Aristotle wrote that mastery of metaphor "is a sign of genius, since a good metaphor implies an intuitive perception of the similarity in the dissimilar."

My uncle James L. Breese, Jr., who was a gifted thermodynamic systems engineer, and whose father, James L. Breese received 136 basic patents in oil burners, wrote to me:

I have always felt that creativity and discovery arise largely from a talent for noticing odd things...Any number of people had seen swinging lamps and chandeliers before Galileo noticed that the period of swing was constant regardless of amplitude (within limits). How many researchers hunched over microscopes in that dusty London lab had impatiently cleaned pollen and fungus from their Petri dishes before Alexander Fleming noticed that fungus was devouring bacteria in the dish? After the discoveries of Faraday and Maxwell, everyone knew that moving a magnet near a wire produced the same effect as moving the wire near the magnet, but it took Einstein to see that this odd occurrence ruled out the notion of absolute motion. Willis Carrier, while waiting for a train to arrive on a foggy night noticed that water drops were forming where wisps of fog swept over rain puddles. He instantly had the counter-intuitive idea that moisture could be extracted from air by passing the air stream through a water spray that was below the dew point temperature of the air. He applied that to solving the problem of high humidity in printing plants and subsequently founding the modern air conditioning industry. Perhaps the greatest noticer of all time was Charles Darwin who saw important differences among plants and animals that countless observers before him had apparently never noticed.

Edmund Fuller, who was a keen observer of American invention, thought a lot about the visionary and creative process of invention. He wrote:

The spontaneous generation of an idea is a factor in mechanical invention as in all other creative processes. A concept, which has not been present in the mind before, simply appears. If the mind is unready or unreceptive, nothing happens (though possibly no authentic concept ever presents itself to such a mind). But if the mind is alert, the concept will set in motion a chain of events which we commonly think of as the processes of invention. Yet it is also true that in the active creative mind, the initial processes of logical reasoning may stimulate and arouse the unconscious resources. There is the real interaction.

3

A VISION AND CREATIVE PROCESS FOR TODAY'S INVENTORS

All great inventors and discoverers have been irresistibly borne onward by faith in the things that could not be seen… To believe and go forward is the key to success and to happiness.
Lilian Whiting

I f you have the patience to read this far, I assume you have created and invented, or you hope to do so soon. None of us is likely to be a Leonardo da Vinci or a Thomas Edison, but we would like to create products that will lead to satisfaction and fortune. Where do you start?

First, You Need a Vision

Virtually all successful business people have been described as visionaries. They could always see a little further over the horizon than anyone else. To develop a great business from a great idea, you need a great vision. A business vision usually starts with a product insight beyond what other people have.

For example, today we take the photocopier for granted, but when that invention's potential market was being researched in the 1950s, very little future was seen for the machine. The earliest photocopies were more expensive than carbon copies, which everyone was using. No major corporation was willing to negotiate the rights to manufacture office machines featuring the carbonless copying process. Therefore, the Xerox Corporation was formed specifically to commercialize the invention. The rest, as they say, is history.

While your invention may be viewed skeptically, even by experts, if you think you have created something that is really good, don't worry. Experts rarely see the merit of new developments, but their advice and assistance is very valuable to carry the initial product concept to a commercially acceptable product. My advice is, no matter what people say, press ahead.

A great vision should carry you beyond attempting to do the ordinary. It should propel you into accomplishing the extraordinary. If you do not start your business and succeed with it, your life will not be complete. Your vision should enthuse you that strongly. You will need this level of positive attitude to get through the rough spots and the lonely interludes of business. Your vision needs to be powerful enough to inject positive energy into your employees. You are going to need them with you every step of the way.

Finding Your Vision

Where do you find your vision? Think in terms of something you feel strongly about or have a passion for. It is best if this something uses tools and techniques with which you are already familiar, or better yet, in which you are an expert. Think about developing something that you yourself would be enthusiastic about buying and using. Let's say you love to fly and you are an electronic engineer. Why not improve instrumentation used in aircraft? Test your vision on yourself. Ask yourself if you would fly a plane using your own designs of instrumentation.

Along the start-up path of your new business, you will encounter many naysayers, critics, and know-nothings. Your best defense against these people is your self-assurance. Your positive attitude will come from your unshakable vision and your understanding about the details that go into making the product and building a business around it. It will also come from a lack of fear of failure.

Many a hobby enthusiast has asked my opinion of the idea of building a business from a hobby. I caution them against it. Do not let a hobby or interesting pastime become your business vision unless it really is the basis for a significant business. Most hobbies do not lend themselves to significant and lucrative businesses. In addition, hobby businesses tend to start in one room of the house and flow to others. The family never likes this, and neither does the Internal Revenue Service. They are very hard on hobbies classified as businesses. Looked at another way, most focused, successful business people would love to have a non-business hobby to get their minds off their business.

People often ask me: "How do I get an innovative product vision when there are so many people developing products?" Basically, you have to crank up to the fullest your sensory and perceptual awareness. For example:

- Look for a simple but attractive improvement to enhance an existing product type. My father made a small fortune after he redesigned wooden bird feeders by making them into more attractive and functional plastic feeders. Despite the simplicity of the product, he received several patents.

- Think about product design in terms of appearance and simplicity. I built a great business making gas detection instruments. My designs were low cost and had fewer switches, buttons, and other controls than the existing devices.
- Look for the connectedness of everything. Engineers, scientists, and academics despair when asked to combine many disparate variables. However, it is perfectly okay for you to combine mechanical and electrical concepts, cooking, kitchen management, and who knows what else to come up with the world's first really good toaster.
- Do experiments, however crude and imperfect, to improve your insights and understanding so that you can move forward with your product design.
- Don't worry about what other people think or say. Keep observing and visualizing. Visualization can be very powerful because it is a key tool for changing consciousness.

Here is the story of how I stumbled on one of my visions and what happened. This is about a high technology business. Its development steps could equally apply to a low tech but innovative venture such as a secret recipe for better pizzas.

The Road from Perceived Need to a Vision to a Company

I was more or less happily plodding along as the partial owner and general manager of a plastics machinery company in New Jersey. I felt that I should develop a new product for quality control during plastics production. It was a gnawing feeling.

I believed that there had to be a way to see the impurities in plastics, called gels.

Gels look like "fish eyes" in tapioca, and to some small extent, they are in all plastic products. They can cause a great deal of frustration by the damage they can cause, ranging from pinhole leaks in milk jugs to runs in stockings. I knew that the market for such an invention was potentially huge.

To those of you who have not searched for gels in plastics, a little background briefing may be in order. When most plastics are cool and solid, they are not very transparent. However, when almost all plastics without colorant or filler are molten, they become transparent. I knew that most plastics processing is done by extrusion or molding machinery. The logical place in the process to detect plastic gels therefore would be within this machinery.

I still did not know how you would see the gels, even in fairly clear molten plastic, because the gels are tiny. A gel is usually smaller than a pinhead, often about the diameter of a human hair, and is floating around in a very hostile environment of high pressures and temperatures, strong chemicals and fumes, and other obstacles. It needed more thought on my part.

I had only a vague idea about how to "see" the gels in molten plastic. An optical approach seemed most promising. What I needed was some sort of very robust probe that would allow for a remote vision on a micro scale into molten plastic. What I was considering was like finding a way to use field glasses to look into a live volcano. It was a challenge, and I set out to solve it.

Then one of those little miracles of inspiration happened. While walking at dawn in the mountainous countryside in upstate New York, I chanced to see dewdrops glittering on a spider's web. That is when it hit me. The light was sparkling from the dewdrops like the sparkles of light from a chandelier. Sunlight shining from the other side of the tiny dewdrops causes them to shine brilliantly as points of light even in the considerable early morning mist. Furthermore, the vibration of the dewdrops in the gentle morning breezes made them shimmer and glitter, so that they stood out even more from their background. This "shimmering" insight would be the key to succeeding in the product development.

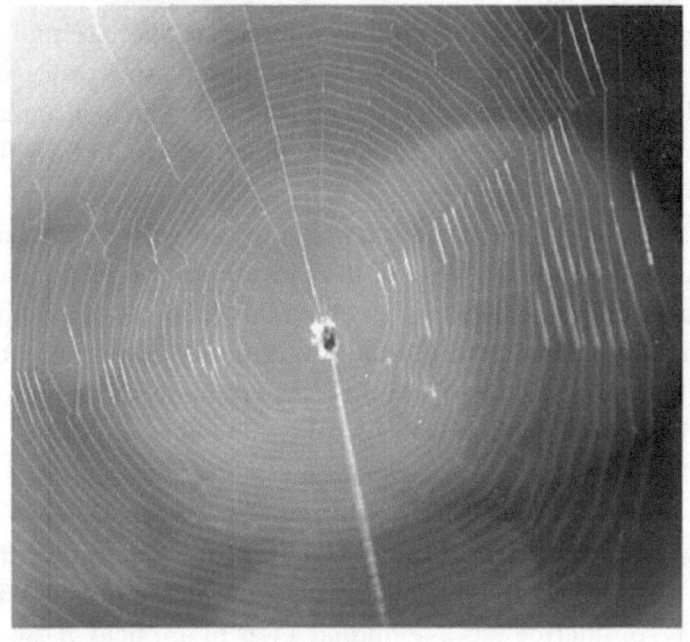

My vision of a polymer gel detector was sparked by
the optics of a spider web in the first rays of the sun.

As an engineer, I saw that an optical concept had presented itself. I could now develop an instrument that would allow tiny impurities to be seen in murky molten plastic. It could not only detect gels but probably count them as well. I was elated by my discovery and anxious to get to work on it.

By the time I returned to my home in New Jersey, I had mapped out both the preliminary design of the product as well as the form and structure of the company I would need to commercialize it. I estimated that at least a million dollars would be needed to finance the new venture. A positive factor was that I managed a small plastics machinery company. I could do some initial new product development within this company, and I was known to be an effective general manager.

I knew I had to develop a working prototype. I would also have to apply for a patent, develop prospective customer interest, and then talk to prospective investors. I had found a potential product concept. Now I had to work towards starting a new company. I knew a major investment would be needed.

Papers were drawn up to form the new company to commercialize my new product, but the new company had no staff, facilities or finance. Yet the idea of a new company seemed to be a *fait accompli*. The details would be attended to when I had a business plan and the company's needs became clearer. Meanwhile, the product development project was carried out within the plastics machinery company that I managed.

Now I settled into the exciting and adventurous work of product development. It took many nights and weekends. For months, I found myself experimenting with various combinations of electronics and optics to "see" the pinhead-size gels in molten plastic.

I needed to illuminate the gels. The most promising technique was one I borrowed from medical endoscope technology. Used by doctors, fiber optic bundles are snaked into the body to see magnified images of interior body parts. I would use a similar process employing a highly protected magnifying lens at the end of a protected fiber optic bundle.

I needed a fiber optic bundle on the other side of the plastic melt flow connected to an intense light source. The result would be that the gels would be illuminated the way the early morning sun had illuminated the dewdrops on the spider's web.

Once the plastic was illuminated, the images of the gels were converted to video. This made automated image analysis possible. Next, I had to figure out how to see the gels moving in a flow channel without the confusion of static artifacts in the channel. The quavering motion of dewdrops makes them stand out much more distinctly from the background. I decided to program the electronics to detect only moving gels so that the static artifacts would not be seen.

Motion detection would come from special video circuitry developed for surveillance and security such as for detecting intruders in an otherwise quiet parking lot. To accomplish this, all of the fiber optics and lens systems had to be specially designed to withstand temperatures of up to 800 degrees F (427 degrees C) and pressures of thousands of pounds per square inch.

I purchased experimental equipment as components: a second-hand video camera and monitor from a war surplus store, fiber optics endoscopes from a medical supply house, and other bits and pieces from wherever they could be found. It was all very crude, but when everything was connected together, it sort-of worked. A little vision, a lot of bargain hunting, and imagination still helped!

My employees and I started simply. We looked at grit particles in water. Then we graduated to grit particles in Jell-O. Finally, we arrived at grit particles in molten plastic. Then we improved our equipment and technique so that we could see the tiny transparent gels in the almost transparent molten plastic. With the video image enhancement, the gels appeared as bright round blobs much like micrographs of blood cells. Eureka! I found it!

To make the gel-imaging instrument useful, it also had to size and count the gels in the polymer flow channel. Therefore, by the time we had a demonstration instrument ready for the first prospective customers, it also incorporated a homemade image analysis computer to size and count the moving gels. That computer used sophisticated artificial intelligence algorithms to substitute for a human in dynamic image analysis although we were only vaguely aware of the new science of AI at that time.

The machine stood almost three feet high and was about a foot and a half wide and a foot and a half deep. It was crammed with electronics. The estimated manufacturing cost of the complete gel detecting and counting instrument was about $30,000. My proposed sales price was $78,500. (Never mind about by what mystical process I arrived at that price except that I do recall it netted a profit margin that is typical for industrial instrumentation. Also, the combination of numbers seemed pleasing.)

By now, a year after the initial product conception, our development team consisted of me as the electronic engineer, a consulting polymer chemist, a mechanical engineer, and a consulting software development group. We were all paid by the plastics machinery company I managed because as a new venture we had not yet raised outside financing.

We needed some market feedback before we invested further in product development. What would they think about my new product in the plastics production plants?

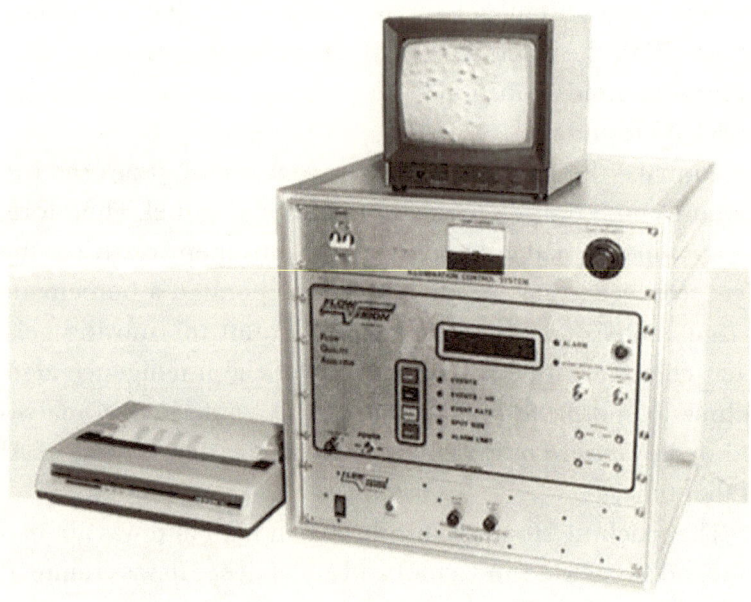

Gel detection instrument developed from the spider web insights. The gels seen on the monitor are in molten plastic and are about the diameter of a human hair.

Our encounters with the market threw us into a prolonged funk. Who would want to buy one of our gel counters at $78,500? Certainly, the average plastics products plant that made cups or plastic bags could not afford it. We wondered where all this hard work had taken us. Perhaps it was just an expensive science project. Maybe my vision had propelled us into uncharted territory where we really did not want to go. We were approaching Christmas, and there were few prospects of tidings of great joy. I thought that I would soon have to shelve the project and concentrate our dwindling resources on our traditional plastics machinery business. It was lower tech, lower profit potential, but a relatively reliable and steady business.

Then the phone rang. I could not believe my ears. The man on the other end of the line said that they would buy an instrument if we could deliver by year-end—about 20 days away. It was Dow Chemical in Texas calling. They had seen a brief description of our product in a trade magazine new product publicity release. Their plastics were shipped in pellet form by the railcar load to the people who make plastic bags and a wide variety of other household products. They had a serious gel problem, and it was affecting the million or so pounds of raw polyethylene plastic they were producing each day.

The catch was that any leftover project funds at Dow were available only if they were spent on goods delivered by year-end. This is common practice in larger companies and government agencies. Our salvation came because someone at Dow did not want to send the funds back to the corporate coffers! We accepted the order without any real hope of completing the instrument on time. You can guess what we were doing on Christmas day.

Word got around. We sold a second instrument to Dow. (We guessed the first one had performed well. Dow generally did not hand out accolades.) Soon after, DuPont bought one to ensure fiber quality control for bulk nylon production. An engineer from Exxon Chemicals appeared one day and the following day there was an order for an instrument to do research analysis of synthetic rubber production. We had found our market or, more accurately, the market had found us. The market was in bulk plastic production, known in the trade as polymer and fiber production.

The true market finding you is a common development. With innovative new products, you very often do not know what the real market segment or customer category will be. Do not be blinded by your initial perception of who the typical customer will be, and allow time for the true market to present itself.

To sum up the product development process, I should refer to my father who was a great inventor and product designer. He said reflected on these matters in an interview with a local newspaper:

My ideas are not guaranteed to make anyone's fortune. But then again—if you're not in a hurry—maybe they will. You ask me how I came up with a bird feeder that made the world beat a path to my door? By doing my work according to these notions:

- **Make sure you satisfy yourself**. Don't let yourself be rushed into doing things you're not proud of.
- **Don't be afraid to fail**. Be willing to work by trial and error. That's the only way you learn.
- **Reach for excellence always**. You can never reach that perfection yourself, but if you try as hard as you can, that's the main thing. That's where happiness lies. And, often, success as well.

THE PROJECTED MIND: YOUR KEY TO CREATIVITY

Imagination is more important than knowledge
Albert Einstein

"For three weeks, the Huygens probe had coasted, dormant, after detaching from the Cassini spacecraft and being sent on its way to Titan. Those of us watching anxiously felt a deep personal connection with the probe. Not only had we worked on the mission for a large part of our careers, but we had developed its systems and instrumentation by putting our minds in its place, to think through how it would function on an alien and largely unknown world." So wrote Ralph Lorenz and Christophe Sotin in the *Scientific American* about their great space adventure.

These space scientists nailed it: to make new theories, new inventions, and other great creations, we have to do better than adjusting existing theories and designs. We must forcefully move our mind beyond the existing thinking about the subject. We must move out of our conscious world and focus our mind in a new place occupied only by the new creation.

Abstract Thinking and Refocusing the Mind

Reduced to its simplest elements, what you are required to do is solve a problem or construct a work of art without a complete set of instructions or without comprehensive data. In a creative process, you are using your imagination to make an appealing or useful whole from a set of components that would not appear to be sufficient or adequate for the job. To do this you need to see beyond mere recollection or simple association. You are projecting the mind's eye to another point in space or time. You are putting your conscious being in an entirely different surrounding environment. One way of looking at this process is that you will be creating a new mind out of your regular mind.

Einstein placed himself in speeding trains, moving clocks and elevators in space. This was more than metaphorical thinking; it was a mind transforming itself to another place. Einstein's strength came from his imagination and creativity. For the most part his mathematics is a precise description of the relationships he discovered rather than the way he arrived at those relationships.

Credit: Roger N. Shepard

Einstein
imagines riding a free-falling elevator

My father invented a number of bird feeders that are the now familiar plastic tubing with metal perches. He started by imagining himself to be a bird on the perch. Then he envisioned the geometry that would be most accommodating to the bird. Only after the birds were satisfied did he select the materials and manufacturing processes to make an attractive and economical product.

He always included testing the feeder designs to see how happy the birds' chirping sounded! I venture to suggest that most would-be inventors or designers of bird feeders would start with an overall design, reduce it to acceptable cost, and finally test in an actual bird-feeding situation. As a consequence, by not following the bird's eye route, they would probably not have an optimum design from the birds' point of view, which would undoubtedly put their product at a disadvantage once they came into competition with my father's feeders.

I myself have three patents and co-patents relating to observing transparent particles and impurities in molten plastic flow and particles in paper pulp slurries. In each case, during the inventing process, I located myself as a microscopic observer in the chemical flow processes where I could see the relationships of particles in the liquid flow and under various illumination geometries. Of course, it was necessary to apply chemical and mechanical engineering, knowledge of materials, and acquaintance with video processing to arrive at a useful design, but the insight came first.

When the inventor comes up with a truly novel idea, he or she has been exploring relationships, patterns, and associations until a productive interplay of ideas, images, and data of all kinds is found. That encouragement signals the brain that the chase is on. The mind is to be projected to a little world encompassed by this project. I call this world imagination space. I know this is what happens because I have done it many times myself.

I am projecting my mind into an external space somewhere outside of my head and my surroundings. This space could be invention space if I am creating new products. It could be a stage in artistic space if I am composing an opera. It could be inside a black hole if I am a physicist working on a new theory. In each case, I am striving to be imaginative and creative by wholly implanting my mind's eye in the space immediately surrounding what is being analyzed and for endless periods of time. I analyze all the combinations of data and search for clues to the breakthrough I am hoping for while unsociably avoiding distractions.

However, none of this imagination is possible without referring to remembered or retrieved information and events. When we compare memorized and reference information with new information, through analogies we can invent solutions to a problem.

Creativity and Brain Structure

Jeff Hawkins has explored this area of memory and prediction in detail in his book *On Intelligence*. Hawkins is a computer engineer and founded the Redwood Neuroscience Institute. More recently, Hawkins founded Numenta, a Bay Area firm that is seeking to commercialize a cortical learning algorithm that mimics the brain's capability to detect complex patterns. In 2014 they announced some commercial products.

Hawkins proposes that our cognitive memory resides in the thin, shell-like neocortex on the top of the brain where it is estimated that there may be as many as one quadrillion synapses available for a lifetime of information storage. His model is that all the brain knows is experiential patterns of sight, sound and smell. There are also temporal patterns showing pattern changes over time.

The memories in the brain are stacked associatively. This means the separate events committed to memory are connected to each other by similar descriptive words or images. Such brain wiring is often called a semantic network.

Sequences of patterns are built that belong together by subject or by sequence of time. New information absorbed in life's journey is "pushed down" in its associated information stack resulting in a more detailed set of patterns. This collection of information is like an automatic filing cabinet.

Hawkins does not believe that creativity is some extraordinary quality that requires high intelligence and giftedness. He defines creativity as making predictions by analogy, and he proposes that it occurs in the entire range of difficulty ranging from hearing a song in a new key to composing a symphony in a brand-new way. The problem with this latter extreme is that to make an entirely novel creation the inventor must think beyond the bounds of what he or she can recall from memory.

Hawkins' analysis is generally in the right direction, but I do not think he takes abstract thinking into sufficient account. Reasoning and abstract thinking are higher level thinking skills. These skills and abilities allow us to think in terms that are neither obvious nor readily referenced. Abstract thinking is a type of creative mind projection.

How the projection of the mind to a uniquely productive imagination space happens is a subject of a lot of current conjecture. One school of thought says that we store much more information in our unconscious mind than our conscious mind. In the intensive imagination and invention process, we search our unconscious memory for clues and ideas and promising ones are resurrected from time to time. One such time we all can relate to is thinking about an unsolved problem just before going to sleep and awaking with the idea in full sound and color. The mind had been rummaging around its archival memories overnight, probably while dreaming.

Freud proposed his concept of the unconscious more than one-hundred years ago, but only recently has his theory received appreciably more scientific credibility. Brain imaging studies confirm that there is indeed an unconscious memory that is referred to under some conditions. These studies however, revealed a potentially troubling fact: Delays between when the unconscious responds and when the brain is outwardly conscious of a response are reported to be in the range of 1/10-1 second.

This caused Nobel Prize winning memory biologist Eric R. Kandel to ask, "If the choice is determined in the brain before we decide to act, where is our free will?" My answer for the invention process is that free will was manifested in the instructions given to the brain's search engine before it dove into the unconscious.

It can also be argued that the human brain has a control element that can be subject to a random event within the mind. Thought patterns can be pursued that could not be previously foreseen if they can be switched randomly to new areas of inquiry or action. If the mind is deviated from a routine thought sequence in the creative process, then it can be launched into highly creative imagination space by a random component of thought.

This aspect of brain activity is philosophical as well as scientific, and as of this writing, practical ways of introducing random elements into the creative thought process have not been conclusively determined. My suggestion is let your mind wander. This seems to happen when I'm hiking. My brain apparently is almost idling, but it seems to mysteriously come up with interesting ideas and solutions I have been seeking.

My feeling is that randomness is definitely involved in many creative discoveries although as Louis Pasteur famously observed "Chance favors the prepared mind." Pasteur also said, "Let me tell you the secret that has led me to my goal. My strength lies solely in my tenacity." One could interpret this as if you look for something long enough and diligently enough you will not need the providential intervention of chance.

Jeff Hawkins asked, "What's different about human intellect?" First, he notes that the human neocortex is larger than other mammals, so that we humans can make a more complex model of the world and make more complex predictions. We can create analytical worlds in an infinitude of specialties or in spaces beyond our surroundings far into the universe. Human intelligence can use thinking which is more detailed, more complex, and which can interrelate more parameters and data than can animals. This is one reason why we can fabricate imaginative visions of things that do not exist in everyday experience.

Cognition and Language

Hawkins says that the reason that our intelligence is much superior to animals is that we have language. While it is generally accepted that many animals recognize dozens of words, there is no evidence that they can learn expansive vocabularies or use grammars. In other words, animals cannot think in or communicate with language.

Hawkins summarizes: "Through language one human can invoke memories and create new juxtapositions of mental objects in another human. Language is pure analogy, and through it we can cause other humans to experience and learn about things they may never actually see."

Noam Chomsky, Professor of Linguistics at MIT, started this line of thought with his proposal that humans are born with a universal grammar. Grammars for spoken languages are quickly learned in early childhood because of the patterns already in the children's minds of the universal grammar. Chomsky speculates that humans fell into language and a universal grammar by a tipping point combining a specialized gene and maybe one or more small physical changes. In his lectures on *Language and Problems of Knowledge,* Chomsky broke some new additional ground:

> Now for some speculation about human evolution. Perhaps at some time hundreds of thousands of years ago, some small change took place, some mutation took place in the cells of prehuman organisms. And for reasons of physics which are not understood, that lead to the representation in the mind/brain of the mechanism of discrete infinity, the basic concept of language and also of the number system. That made it possible to think, in our sense of thinking. *So now humans— or prehumans— could go beyond just reaching out to stimuli and could construct complex structures out of the world of their experience, and now, the world of their imagination.* Perhaps that was the origin of human language. (emphasis mine)

The gene has been identified which is relevant to the human ability to develop language. It is FOXP2. This language gene was first isolated in studies of three generations of an English family who had extreme speech and language disorders and who did not have this gene. It is located on chromosome 7.

Other versions, which apparently do not elicit language capability, have been identified in songbirds, mice, chimpanzees, gorillas and other animals.

However, the protein encoded by the gene found in humans is slightly different from the protein it encodes in other animals, and this may be why humans have unique and extensive language capabilities. The field of genetics and language is rapidly changing, and new developments and revisions of theories can be expected frequently.

Integral with the use of language in the world of imagination, is the dialoguing process, often with one's self. Arthur Koestler described this process as, "Creative activity could be described as a type of learning process where teacher and pupil are in the same individual." Einstein said: "The important thing is not to stop questioning." This is an important tie-in between language and extreme creativity and innovation. Only man asks questions. His imagination poses potential answers.

When I am inventing, I am in a constant dialog with myself:

Well, let me see. If I want a support for this wide, thick plank, moving it forward, I will need something flat that keeps the plank off the ground.

Hmm. How about some sort of disk?

Seems promising…How does it connect to the plank?

Put a hole through the center and have a big spike fix it to the side of the plank.

Well, wouldn't it be easier to put two on the plank—one on each side—and push the plank with a handle on the rear edge?

Or better still, how about two pairs of rotary support disks on the plank? It would be more stable.

I built the first prototype using stock wood from the shop and a few tools. It squeaks when pushed. The disks wobble uncontrollably, and the contraption is generally unsatisfying. So it is back to the drawing board:

Couldn't we design a cylindrical fitting...how about a cup shape?...With a big round hole in the center...it could slide over a round support rod with grease between them. This fitting could be press fitted into the disk.

Hallelujah! I just invented the wheel, hub, and axle. Of course, I could have done this without language, but for me the process would have been much more tedious.

I also go through an interactive development process like the one above when I am interacting with Google in a search for ideas and components to support a possible invention. Google is just a passive recipient of queries in this dialogue. Someday, Google or other search engines will ask questions in response to inquiries in order to speed up and sharpen the development process.

Recently, while at the shop, I received a visit from two bright kids about 12 years old. Out of curiosity, and to gain insights about the invention process, I decided to conduct a little experiment. I showed them the workbench where there was an assortment of small empty bottled water bottles and screw on caps, empty soft drink cans, and various sizes of rings, rods, and fittings. I asked them to develop the wheel and without a lot of conversation about it as they experimented. I was surprised at how long they struggled with fitting parts together and trying to evolve a wheel. Not using language or dialoguing was clearly a great handicap to their creative process.

When I asked the kids what they would do next if they had more time, one replied that he would do further experimentation after searching around for more experimental parts, and the other said he would research ideas on the Web.

There is much debate about whether the creative process starts with images or language. Antonio Damasio in the classic analysis of cognitive thinking, *Descartes' Error,* says:

> It is often said that thought is made of much more than just images, that it is made also of words and abstract symbols…In this regard it is interesting to observe that some insightful mathematicians and physicists describe their thinking as dominated by images.

While language is a common way for creative people to force their minds into creative spaces that normal thinking would not bring them to, and it will become more important as research is done using the Internet, there are other ways to do this.

Temple Grandin says, "I think in pictures." She is autistic, a lifelong student of the learning process, and an equipment designer for the livestock industry. Grandin received a bachelor's degree in psychology and master's and doctor's degrees in animal science. She is professor of animal science at Colorado State University, a bestselling author and consultant to the livestock industry.

Grandin writes, "When I invent things, I do not use language. Some other people think in vividly detailed pictures, but most think in a combination of words and vague, generalized pictures." She describes her visualization abilities projecting her mind as follows:

Early in my career I used a camera to help give me the animals' perspective as they walked through a chute for their veterinary treatment. I would kneel down and take pictures through the chute from the cow's eye level...They helped me figure out why the animals refused to go in one chute but willingly walked through another.

On the other hand, my thinking is non-stop verbal dialog. I think Grandin would accept this to be logical because even though I am trained as an engineer, I have always been verbal. I was the features editor of the college newspaper in my sophomore year, and I scored higher in the verbal scholastic aptitude tests than in the mathematical abilities tests.

My father on the other hand, dropped out of Harvard in his senior year and became a successful inventor and small business entrepreneur. He was never very verbal and in fact was probably dyslexic. He liked to think through mechanical ideas with me by talking in terms a stream of drawings he would produce on a sequence of 3" X 5" cards. They showed parts coming together and pulling apart with occasional labels, dimensions, and notes.

He would always project himself into the invention space at the beginning of these thought sessions with me by saying something like: "Imagine you are a gear on this shaft and you engage with another gear located over there..."

The Emotion Factor

Emotion is very important in high-level thinking, but its exact relationship to the thinking process has not been defined or detailed exactly.

Marvin Minsky, Professor Emeritus of Electrical Engineering and Computer Science at the Massachusetts Institute of Technology and cofounder of the MIT Artificial Intelligence Laboratory, in his book *The Emotion Machine*, explores emotion as an outcome of several states of mind. He writes that our various emotional states result from turning certain mental resources on and off.

Minsky adds that we are programmed from birth with responses to anger, hunger, fear and thirst. As we learn and grow, we add mental states of a more intellectual nature starting with learned reactions, progressing to deliberative thinking, reflective thinking, self-reflective thinking and winding up at self-conscious emotions.

On the other hand, a heightened emotional environment may cause the thinking person to switch into a higher or lower level of mental activity instead of the other way around. This may be particularly true for the creative thinker. Again, Antonio Damasio writes:

The cognitive mode, which accompanies a feeling of elation, permits the rapid generation of multiple images such that the associative process is richer and associations are made to a larger variety of cues available in the images under scrutiny. The images are not intended for long. The ensuing wealth promotes ease of inference, which may become over inclusive. This cognitive mode is accompanied by enhancement of motor efficiency and even distribution, as well as increase in appetite and exploratory behaviors. The extreme of this cognitive mode can be found in manic states.

One emotional influence is the socioeconomic pressures of a given era. Times of tension and danger often seem to foster creative and inventive efforts. The monumental creativity of the artists of the warring states of the Renaissance and the incredible scientific developments made during World War II are classic examples. Was the creativity enhanced because there was a top-down pressure on creative thinkers to work even more diligently on their projects and possibly save civilization?

Or was it because the pressure of the times somehow modified the brain chemistry of the creative people forcing them into an involuntary manic mode much of the time? The influence of chemicals ranging from alcohol to narcotics to alter the imbalance of neurochemicals including serotonin and dopamine has been noted from ancient Greek times to the present. Many creative people such as writers feel that immensely creative states of mind are reached when the mind is not in a normal equilibrium state. Everybody has heard of the swings from writer's block to great bursts of creativity.

The *Wall Street Journal* reported that Ritalin and other drugs for attention-deficit hyperactivity disorder (ADHD) have helped many children improve their focus and behavior, but their creativity and drive may have been dulled. Among the very creative people thought to have had ADHD include Thomas Edison, Albert Einstein, Salvador Dali, Wolfgang Mozart, and Winston Churchill. Ritalin's generic name is methylphenidate, and how it affects ADHD patients is not well understood. One theory is that ADHD is caused by a dopamine imbalance in the brain, and that this increase is partially blocked by the methylphenidate.

It has also been reported that artistic brilliance and a dazzling memory are related to autism and other developmental disorders. Temple Grandin puts it this way:

It is likely that genius is an abnormality. If the genes that cause autism and other disorders such as manic-depression were eliminated, the world might be left to boring conformists with few creative ideas. The interacting cluster of genes that causes autism, manic depression, and schizophrenia probably has a beneficial effect in small doses.

Then Grandin explores another emotional element that is very intriguing:

Similarly, being childlike may have helped me be creative. In his book *Creating Minds*, Howard Gardner outlined the creative lives of seven great twentieth-century thinkers, including Einstein, Picasso, and T.S. Eliot. One common denominator was a childlike quality. Gardner describes Einstein as returning to the conceptual world of the child, and says he was not hampered by the conventional paradigms of physics.

I think as we get older almost all of us yearn for the carefree times of childhood. Despite the many problems that hemmed in almost every child, the child still has the almost naïve capability of unfettered imagination. Some people, very few, keep this imaginative ability through adulthood. Their imaginings leads to inventions, art, designs and explorations of many frontiers never seen before. Emotion is part of this creative formula, and perhaps the emotional element is what is hardest to reconcile in equating the human mind to an advanced computer or an artificial intelligence machine. Did you ever see a computer cry?

Imagination gets us beyond the here and now. It gives us the ability to ask questions in a new spirit of discovery. It facilitates seeing ahead and exploring the best way to go. This is an essential step to go from imagination to creativity to invention. This puts us in a class distinct from very smart animals and super intelligent computers.

5

THE AMAZING BRAIN: HOW IT WORKS IN THE CREATIVE PROCESS

We are not who we are simply because we think. We are who we are because we can remember what we thought about.
Larry Squire and Eric Kandel, *Memory: From Mind to Molecules*

Thinking about thinking has always fascinated me. It started when I was in graduate school at MIT in the mid-'60s where I did some original analysis of artificial intelligence. I focused on the idea that the mind assimilates information by association with similar information it already has. By projecting from a sequence of correlated patterns, the mind decides what to do next. I called this associative inference. It is a simple idea but was not widely discussed at the time.

Evolution has brought us to the point where we can remember what we thought about, and with our present mental equipment, we can also imagine, create and invent. Will we humans continue to dominate the creative process or will super computers assume this function? It is worthwhile looking at the thinking part of the brain to see where we fit compared to the evolution of computers.

The Basics of the Thinking Brain

Developing the brain to its current level of sophistication in human beings has taken billions of years beginning with the advent of life itself. The earliest organisms that were around over a billion years ago came preprogrammed by the software called DNA. This produced genetically inherited information which changed and improved from generation to generation. Some genes in the DNA contained the instructions for manufacturing the organism, and some gave it automatic direction to follow up on clues about food, presence of light, temperature and the other environmental essentials of micro life.

Fish, birds and other non-mammalian creeping, swimming and flapping creatures were packaged with a memory-based improvement to deal with moment-by-moment changes in their environment. This was the allocortex, which allowed simple responses to various observed situations, but not cognitive intelligence at the level of, say, a dog or cat.

A big step in the development of the brain was the emergence of mammals. The timing of their very beginnings is fuzzy, but they diversified quickly and flourished after the extinction of the dinosaurs sixty-five million years ago. The mammal brain includes an addition to the allocortex called the neocortex. It is the outer layer of the upper hemisphere of the brain and consists of six layers. Its functions include sensory perception, the ability to learn with behavioral flexibility, motor commands, conscious thought, and, in humans, language. In humans, the neocortex is 90% of the cerebral cortex up from a very minor percentage in simple mammals.

Another important part of the brain, which makes it a superb thinking machine, is the prefrontal cortex behind the forehead. It relates current perceptions to memories of past experiences in order to provide an executive and planning function for judging priorities and future planning. It plays a critical role in the regulation of emotion, and it is the storage area for short-term memory. The prefrontal cortex is involved with decision-making, problem solving, and integration of ideas. Other mammals, particularly primates such as monkeys, have prefrontal cortices but not as developed.

Also important is the hippocampus, a small component in the lower central brain, which manages information flow to the long-term memory. It also is important for spatial navigation such as finding your way around the room.

The fundamental component of information storage in the neocortex is the neuron, a biological cell. The neurons have information gates called synapses to connect by electrochemical signaling to other neurons. The neocortex nodes are the intersection of billions of interconnected synapses resulting in a virtually solid mass of ultra fine wiring. Francis Crick, co-discoverer of DNA, hypothesized that these brain cells represent the mind in the cognitive sense, no more or less. Spread around in that gooey mass are all your memories, skills, and the pathways interconnecting them.

How big is this cerebral mass in modern computer terms? The neocortex has about 100 billion ($1X10^{11}$) neurons, which are about the same as the 100 billion bytes (100 gigabytes) of a typical recent model personal computer. Not encouraging so far in view of the fact that many PC memories have to have their storage capacity increased. Next, consider synapses, the interconnections between neurons. Estimates here vary widely, but a middle ground estimate could be 5,000 ($5X10^3$) synaptic connections per neuron. This totals half a quadrillion ($5X10^{14}$) synapses. This is about half a billion synapses per cubic millimeter—an unimaginably compact packing density.

The number of synapses is the most important number because it is a rough approximation of the brain's information recognition capacity. The bottom line is that the human brain has enough memory capacity to accommodate most reasoning and cognitive tasks.

An interesting indicator of total brain activity is that the brain accounts for about 20% of the body's energy consumption while it only accounts for about 2% of body weight. It consumes about fifteen watts of power, which is much less than a digital computer processing the equivalent amount of data.

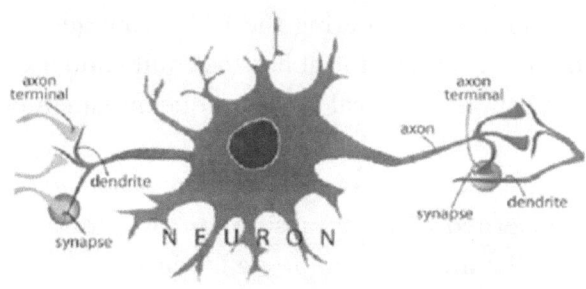

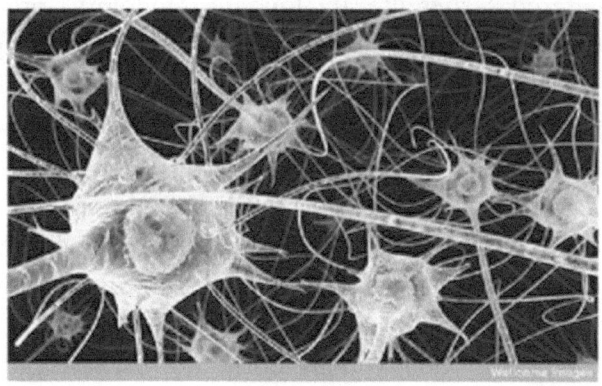

Neurons, dendrites, axons and synapses

Much of the rest of this chapter will be devoted to teasing out a schema of the mind by which we can visualize and describe intelligence. We will explore how information is stored and retrieved in those 100 billion neurons and over one quadrillion synapses. This is still a very lively area of current research, and the interested reader is advised to stay current by following publications that report on cognitive science.

DNA programming is done both by "nature" and by "nurture." The nature part is incorporated within the philosophy of nativism, which argues that certain abilities are included in the brain at birth. These abilities would be included in the genes and modified by altering the DNA from generation to generation. For example, the ability to acquire and use spoken language effectively is critical to the unique human intellectual capability.

The nurture part, also referred to as the *tabula rasa* or blank slate, is concerned with the absorption of information and abilities into the human brain over a lifetime.

Eric Kandel revealed that synaptic connections are weakened or strengthened with experience and that the neurons are not altered. Genes alter the long-term memory stored data by responding to information forwarded from the short-term memory. Input from gene activation proteins called CREB-1 are balanced against the input from gene suppression proteins called CREB-2. This is one reason why you do not remember everything you have learned. He has pointed out that "memory reinvents itself. Every time you remember something, you modify it a little bit." He adds that memory is always a mixture of images, feelings, words, and other percepts.

From Day One when you are born and first see light, your brain is an almost completely empty filing cabinet. With each event, you enter information into the filing cabinet where it is filed and indexed by some process unknown to you. The logical thing for your brain to do is file by subject with reference notations so that the contents in any one file can be referenced to the contents in all other files.

The connection from one to another is by a fine fiber extension called an axon. It connects to the synapse connection at another neuron. With about seven-thousand synaptic connections per neuron, there are sufficient connection options to describe the subject in terms of size, shape, color, sound, odor, taste, name, and relationships with beings and places.

The first perception you will have after birth probably is your mother. With further sensory input the "mother" file is connected to files such as sounds associated with her (spoken name, etc.), visual (appearance, etc.), places where she is found, other people she knows, and so on. You develop a similar grouping of files for "father" interconnected by common files such as "family" and "our house." You are the central connection point to all those files, and you can add to them or read from them at will. In similar fashion, you add Fido, the family dog, your neighborhood, the weather and other important things in your life.

Piece by piece, you are building your personal cross-referenced file which taken together describes your environment and accumulated experiences. As the days, weeks, months, years, and decades pass, you keep adding files which accumulate additional information as your horizons constantly widen: new family in your house and living elsewhere; friends and groups of friends; towns, cities, states, countries; objects described by color, sound, location and so on. Your nearly instant access to this unique universe of files to give you a here-and-now picture of you in your environment. It represents a key part of your consciousness.

In the brain, we go from detail, to aspect, to percept by going from neurons to nodes to networks of nodes. Virtually all neurons are interconnected to other neurons in the brain by the synapses. When we compare percepts, such as your dog to another dog, your mind does a cognitive matching process, which is a form of associative inference.

Notice that with the neurons cross-referenced by thousands of links per neutron, you can do a search of any topic, and pull up any or all of your stored information related to the inquiry. Friends who have brown hair: this sets off a Google-like search using "brown", "hair", and "friends" as related file linkages. You bring a new person into your memory who you are not sure is a friend, but they have brown hair and could be matched up with all other related information in your memory. Then they are wired in as a friend with brown hair or possibly a stranger initially mistaken as a friend with brown hair. This associative inference process is also referred to as thinking in analogies and pattern matching. As part of the creative process, you can make predictions by analogies comparing patterns or comparing the same pattern with itself from an earlier time.

In order to put the whole picture together, we have to see what is known about how the brain projects into the future based on stored information from past experience. Daniel Schacter, Professor of Psychology at Harvard University, and collaborators did important research in 2008 where they found a cognitive process they called "simulation." It allows us to imagine, plan for, and predict possible future events. They found that we can place ourselves in a hypothetical scenario and explore possible outcomes. This ability does not appear to be present in animals or computers.

The hippocampus which I cited earlier for its role in managing information flow to and from the long term memory also seems to be crucial to the reintegration of details in order to recollect a past event and to recombine details into a simulation of a specific future event. Current research supports the role of the hippocampus in supporting relational memory processes.

Incidentally, it is interesting to speculate about how animals think. Schacter reports that there are compelling experimental demonstrations that cast doubt on the claim for human uniqueness. He cites studies of jays and rats, which exhibit strong planning-like behavior in such areas as storing and retrieving food. Anyone who has a pet dog is likely to agree. However, Schacter is not certain if the debates about mental time travel in animals will ever be settled because we have not found a way to communicate with animals about their thoughts!

fMRI Brain Scanning and Associative Inference

The studies by Schacter and others drew from a combination of neuroimaging, neuropsychological, and cognitive studies to arrive at models of thought processes. Neuroimaging, especially functional magnetic resonance imaging (fMRI), has lead to a continuing stream of breakthrough studies to better understand what is going on in the brain. Its colorful images of the inner workings of the brain are beginning to appear in television science shows and popular magazines.

fMRI is a specialized application of magnetic resonance imaging (MRI) which has been used for some time in clinical studies of patients' organs and structures. In MRI studies, the patient is placed in a powerful cylindrical magnet the size of a small room. This magnet creates a field roughly 10,000 times stronger than the earth's magnetic field. Also, radio waves are imposed in the area of analysis. The combination of the radio waves and the magnetic field causes atoms and molecules to emit weak signals.

These faint signals can be detected by sophisticated equipment indicating chemical differences in cross sections of the brain tissue. The whole area of analysis is scanned in thousands of small steps and the resultant complex mass of data is analyzed by computer to look for patterns. Hydrogen atoms, which are part of the molecules of water, carbohydrates, proteins and sugars, show strong signals and thus give clear images. Therefore, fatty tissue, which has many hydrogen atoms, looks bright and bone, which has few hydrogen atoms, looks dark.

fMRI focuses on the neural activity in the brain to see which parts of the brain are active at any time. It detects releases of oxygen to the activated neurons. While at first glance, fMRI's basic limitations might look overwhelming—such as a resolution of several millimeters versus the fractional millimeter size of neurons themselves, and the response time of several seconds versus the fraction of a second response time of neurons—using chemical enhancement techniques, statistical analysis, and various computerized models very useable data can be collected and displayed. Still, the number of studies is limited, due in part to the very high cost of the fMRI equipment, computer systems, and specialists involved and in part due to the tedious experience endured by the human "subjects."

A path-breaking study using fMRI of how much the brain associates related conceptual knowledge, titled "Predicting Human Brain Activity Associated with the Meanings of Nouns," was done by Tom Mitchell, Professor of AI and Machine Learning at Carnegie Mellon University, and associates. Their study was conceived to build upon brain imaging studies that have shown different patterns of neural activation are associated with thinking about different categories of pictures and words such as tools, buildings and animals. fMRI measured neuronal activity for volunteers who were asked to think about nouns such as "celery" and verbs such as "eat" and "taste" (related to celery) and "ride" (not related to celery).

Responses for sets of words that are related and for sets of words that are not related were detected as bright areas in the brain by fMRI scans. These were compared to linguistic data bank statistics of the probability of co-occurrences of different words to be found together in thousands of texts. In simplified terms, Mitchell and his associates found by fMRI that you think of "chew" with a greater degree of association with "celery" than the relative association when you think of "celery" and "ride." The degree of association of "celery" and "chew" corresponds semi quantitatively to the degree of association between "celery" and "chew" computed after reading millions of texts where those words co-occurred. The brain's verbal associations seem to agree with verbal associations that occur in printed literature.

Google developed the word text data bank with frequency of co-occurrences referenced in Mitchell's experiment, and it is available to researchers through the University of Pennsylvania. Approximately a trillion words are in the data bank. This pattern development becomes much more statistically valid when these unimaginably large numbers are used, and it will become an important concept later in this book as pattern recognition in on-line searches is discussed.

Mitchell and his collaborators summarize:

These results establish a direct, predictive relationship between the statistics of word co-occurrence in text and the neural activation associated with thinking about word meaning...This is a natural progression as the field moves from pretheoretical cataloging of data toward development of computational models and the beginnings of a theory of neural representations.

They conclude that while more research is required in predicting fMRI neural activity, they are on their way to understanding the neural representations of meaning.

Their computer model has been tweaked to interpret sentences. It shows similar brain scan patterns for a bilingual person thinking of a noun in either language. Google has used similar analytical techniques involving word co-occurrences as part of their on-the-fly spell-checker and in their language translator.

Here we see an emerging paradigm of the human mind: apparently using associative inference to search for analogies between the brain's stored information and incoming information. This process is similar to pursuing ideas developed using on-line search programs and community data programs using queries and locators based on associative inference. This will be an important consideration as we try to see how the creative human mind can work most effectively with the computer cloud.

It should be evident by now that we cannot really think about an idea for which the brain has no significant related reference information. This is a basic reason why geniuses cannot produce really great ideas or creations simply because they have a very high IQ. Through years of study and piecing together bits of information, they accumulate a cache of information, which is necessary to find a solution to a problem. To see a relationship as yet unseen by other researchers, the genius thinker must project his or her mind to a space that has a perspective which uniquely encloses and defines the problem unfettered by conventional thinking about the topic.

Consciousness and Creativity

It could be argued that consciousness at this moment is the node where you are now connected by synaptic links to all other description terms of your immense associative memory. This includes all stored data (long-term memory) plus current sensory inputs (short-term memory and incoming information). As Eric Kandel succinctly put it, "You are your synapses." In non-neuronal terms, consciousness could be described as where you are, what you are doing and, in a particular sense, who you are.

If consciousness were the only achievement of an advanced mammalian memory, then we humans would only be the equals of dogs or elephants. What really sets humans apart from animals is our thinking and reasoning capability, and this is most facilitated by language and abstract thinking.

Some of the language relationships in the human brain were described in the Mitchell study noted earlier. Dialoguing with one's self, or self-talking, allows us to expand our inputs from information stored in our own memory as well as in the surrounding environment by searching inquiry:

- Why is the sky blue?
- How big is that house?
- How far is it from Chicago to New York?
- Why can't I see electrons flowing through a wire?

Furthermore, it is generally acknowledged that there is an emotional element to creativity. Increased emotion can elevate the mind to higher levels of thought. Emotion is released after the Eureka! discovery moment. Computers of any kind, as far as we know, do not experience this emotion or any other.

From this starting point, we can arrive at the three modes of the intellectually stimulated mind: imagination, creativity and invention. With imagination, we can see beyond mere recollection or simple association. It gets us beyond the here and now. The mind's eye is projected to another point in space or time.

A mental picture or vision is created without complete data or information.

Creativity moves forward from imagination to solving a problem or creating a work of art without a complete set of instructions or recipe. Creativity uses imagination to make an appealing or useful whole from a set of components that would not appear to be useful for the job.

Invention carries this process forward still farther. It cycles through imagination, creativity, and experimentation powered by persistence to develop a new product or process.

6

HOW TO INVENT WITH GOOGLE

The Ultimate Search Engine is as smart as people—or smarter.
Larry Page, Google Cofounder

Any help we can get in our discovery process leading to an invention is welcome. Something that is as private as a library but faster, more comprehensive, and accessible from home. Such a resource is Google using both its search function and its patents site.

How do you know when to start creating with Google and how to go about it? It is much easier when you have an actual creative design challenge. It helps greatly to have the motivation of getting useful information tied to a real goal.

This happened to me when I decided to make a simple instrument to detect common household fumes known technically as volatile organic compounds. I started with a cheap Japanese sensor designed to detect patients' bad breath in dental offices. It worked in my application, but sales were not exciting. One day a customer reported that it worked well for detecting ozone from photocopy machines and that a huge market awaited for photocopy machine repairmen who install ozone filters.

Meantime, however, the Japanese supplier discontinued the product because his dental market fizzled. I had to scurry to find a new sensor and hastily searched the recently established Internet for a new manufacturer of the same kind of sensor (heated semiconductor) used in my design. I found a startup company in Switzerland that probably could only have been found on the Internet. I also had to research circuit design sites on the Internet to find a way to use this sensor for ozone instead of household fumes. The integrated circuit (IC) chip manufacturers' Web sites had the circuit design guidance required.

From there it was a question of making prototypes, having customers try them, redesigning based on their feedback, and eventually finishing up with a design ready for manufacturing and sales. While I never sought to patent the design because of the cost involved for only a two-person startup, my fledgling company grew every year with little competition, and I wound up selling it successfully to a larger company in its industry.

While I could have done all the component search and product design without cruising around in Google or similar Web-based information sources, the time to production could have taken years, not months.

The immensity of the computer clouds offers us all kinds of creative opportunities and resources. The question is how best to optimize human and artificial intelligence in the Age of Google. What do I do best? What does the collective intelligence do best? What do the computer clouds do best? And how can I make the most effective use of their combination?

Beginning the Search and Inventing Process

There probably is no one way to be imaginative and produce a great innovation or creation. Like any other project in life, it is most effective to have a general plan in mind just as if you were setting about to build a house or travel to a distant city. First, spot the need your invention will satisfy. Next, determine what product or service needs to be invented to satisfy that need. Then identify the obstacles to the product's proper functioning or in offering a service.

The vision and invention process starts in your mind, and when it is in a state that can be discussed, you generally should engage other minds when you feel that there is no danger of giving away the whole recipe. This could be a group of friends or professional associates, or perhaps a special affinity group or wiki. They can add to your vision and list of things to do. They can point out probable dead avenues in your approach.

In creating and inventing, you are trying to control the information flow to your advantage. If possible, you should always have an off-line model to experiment with, free of computer communications to the clouds. This would include your notes and drawings, an engineering prototype, circuit boards, chemical formulations and so forth that collectively contain more of the essence of the ideas than your Internet correspondence.

A key part of the creative and imaginative thinking process is "persistent thinking." This is reflecting on your present design solution and comparing it to everything you recall from memory, with what your support group advises you, with what you have read and heard, and with what you can find out from the search engines. This is a repetitive process, constantly comparing the model of the invention in your mind with proposed design variations.

Eventually you will find a conceptual solution to the problem. As time passes, you cycle through the process above mining for new information, experimenting with physical models of the latest concept, and comparing ideas with other people. You should be alert for unexpected clues to understanding the

MEGAMINDS

phenomenon or solving the problem. Eventually an optimal or at least suitable solution will present itself.

Throughout this long and often lonely journey, you should never lose your imagination and curiosity. You should not lose the creative person's enthusiasm, sense of purpose, propensity to ask profound questions and resolution to think boldly. This is what the masters like Einstein and Edison taught us.

Do not get discouraged or overly distracted. You are entering the phase of the mad inventor, and there is nothing anyone can do but put up with it. When your mind starts to wander, either make a new effort to focus your thinking or, failing that, reboot. Go for a long walk. Give your mind an opportunity to reset its programming. Suddenly a new answer may come because your mind has been working on the problem subliminally all along. Write down the insights before they disappear!

Inevitably, your emotional ups of creativity will be offset by emotional downs that can even lead to depression. You must stoke your ego and keep thinking about what fascinates you. When I am in a mental block, often I will cruise the Internet in my general area of interest and see if I meet my cyber muse. New ideas and insights may spring forth.

Using a Search Engine

A comprehensive search engine like Google can make creativity and invention much more efficient and insightful: Often its use leads to serendipitous results for first-time users trying to find solutions to new problems.

Enter a few search terms and the requested package of knowledge appears on your screen. Apparent success (a "Google solution") has probably stalled many promising searches for a required innovation because the person searching using

76

Google did not know when they had reached a viable solution or when they should change their search strategy.

People tell me that they cannot find the important and novel solutions they are seeking by Google searches. This happens if searches are done superficially. To get where you want to go in Google, you must devote considerable thought to what terms (places, things, people and other references) will link you in to the sites that you will want to visit.

My searches may cycle through over dozens of iterations as I try to garner clues from one search about what key words to use on the next one. After I try various combinations of key words and reading selected content of prior searches, the mother lode suddenly will appear. Each search will produce a dozen or more pages with ten or so suggested sites per page.

I find that usually the first two or three pages are by far the most productive. Your most effective searches will build upon key words and ideas learned earlier. Searches will keep building upon prior searches.

Relevance and perspective are often gained from Wikipedia sites found in Internet searches. Wikipedia has provided unexpectedly important insights and data for my development projects.

Google is often considered just a very convenient, infinitely large, and very up-to-date reference library. In the classical sense, it is, but for the imaginative, curious thinker, it can be much more. That is because a very fast stream of data, information, references, and concepts are being exposed to the searcher. New associations and references will occur to the searcher that never would have happened if the searcher was just plodding through libraries.

The Google searcher is flying through a time warp universe defined by all the works of all time that might be relevant to the search and journey. Some people will sleep through this journey and others will absorb all the potentially relevant information along the way.

Six Steps for Innovating and Inventing Using Google

What is the optimum way to make use of Google's comprehensiveness and speed? In addition to following the basics of any well-designed search project, there are at least six important steps as described below.

1 – Narrow the search and find prior art and solutions. This is the time to vent your naïve enthusiasm and surf all around in Google and other web sources. Get a sense of the lay of the land. Which direction do you want to be headed? Has somebody already invented what you were seeking to invent? If not, why not? Have some obstacles unforeseen by you been discovered by others? Can you still envision a solution, not yet offered on the Web, which is worth pursuing? What are the key technical or design problems that need to be overcome? Can they successfully be addressed by your resources and ingenuity?

In addition, if you are interested in patent issues at an early stage in your project, Google http://www.google.com/advanced_patent_search and other Web services offer efficient patent searches.

2 – Postulate a design or system solution. From what you have now learned from the Web, can you make sketches of your imagined creation? For example if it is mostly mechanical design, can you envision the placement of motors, gears, shafts, controls and so on? If it is electronic, can you draw a block diagram and logic flow chart for which specific circuits, IC chips and software can be dropped in later?

3 – Look for design elements on the Web. Check the Web and other sources for suppliers, parts, and ingredients for your creation. Sometimes seeing different parts or ingredients than you originally had in mind will cause you to improve your design. It never hurts to buy some key parts and start physical experimentation, if for no other reason than to further focus your mind on the essence of the solution.

4 – Design and redesign. Perfecting the smooth response of a mechanical mechanism, rewriting computer code, or experimenting with electrical circuit variations can be done and redone almost infinitely. This is part of perpetual creation and invention. This is also part of the relentless quest for full understanding of the process and perfection in action. When inventing interactively with Google, iterate back and forth keeping in mind both the big picture and the details. This is combining you and the computer clouds for greatest creativity.

You can tweak your design too much or too little. It is important to know when to stop tweaking your invention. Most people tend to err on the side of premature product introduction. I have found with my inventions that in retrospect it would have been better to not worry so much about being first to market and better to have worried about reliable product performance.

Preproduction prototypes can be shown to critical observers in your shop or lab, and if those reviews go well, trusted customers can try out the preproduction prototypes. The revised US patent law gives US inventors a year to technically evaluate and get market response to their invention in a public setting before filing the final application. There are similar provisions in many other countries.

5 – Find other people, companies and projects with similar interests with whom you can communicate. This partially refers to the collective intelligence groups and wikis discussed in the previous chapter. These groups are not always helpful in the detailed design process, but they can be helpful in giving feedback about a market application you have in mind, offering novel suggestions about different ways to look at your project, or having some technical data that only a specialized collaboration group would know. You can do much of this on the Web.

It often happens that you will have chance encounters in cyberspace as well as individual people ferreted out on the Web who may be of special interest to you such as professors, writers, skilled tradesmen, and software writers. Of course not all of your new acquaintances may turn out to be as friendly or useful as they first appear—some may well turn out to be jealous or competitive—but I think it is best to get involved with other people at this point and weed out the undesirable ones as you go along.

6 – Organize notes and source material. In the past, note taking, indexing and filing consumed disproportionate amounts of project time. While this essentially manual process will never be eliminated, Web services such as Google, plus similar information access technology for use on data stored in your computer make it easier to manage a database.

By noting search terms that work particularly responsively in Google, I have defined my personal space in Google search. By knowing what wiki or blog is closest to my interests, I have defined a link to groups of collective intelligence. By using a search engine to find things in the impossibly large and growing database in my personal computer, such as Windows "Search programs and files," my personal database is more responsive to queries and, hopefully, no data or insights are lost. Particularly important is that crucial data such as pages from Wikipedia can be migrated to my home computer database.

In addition to gathering information, Google can also channel the creative mind into the unique imagination space that could not be found by simple daydreaming. Seeing what potential suppliers of parts and services do, and seeing the features of similar product designs, can give you new ideas where to fine tune your focus. If this approach is not productive, give your mind free rein to wander again. This is stepping back and searching anew for the way forward.

While you may be very pleased with the ease of organization and use of data stored in your computer, you should not forget to record your original research notes in a "lab notebook." This should have sewn-in pages and the notes and sketches should be written in ink. Each page should be dated and each block of information read, understood and signed by an observer knowledgeable about the subject matter.

Your lab notebook is the primary legal basis of what you observed, thought about, and when. It is often the major document used in the legal defense of the validity of a patent.

A Design Example

In the pre-Google days, you had to start with parts something like you would be designing. Let us say that you wanted to design a super high-strength rivet for aircraft. You would find or buy a dozen different rivets, set them out on the workbench, examine them, make some sketches of a design concept on your pad, and then set up the machine tools to make a first design prototype of your design.

Now, with Google, you might type in the search terms "design of aircraft rivets" and (as of this writing) in less than a second Google would offer a number of candidate pages. The very top one, http://www.nationalrivet.com/

custom-engineered-fasteners.htm will bring you to a very informative rivet design website featuring drawings of all kinds of rivets. You could study these rivet designs to get your mind programmed to think about designing your own rivet.

You might say, however, "What's so difficult about visualizing rivets?" It gets a lot deeper from here. I tried Googling "riveting in space" and only found space photos that were riveting (attention getting). I tried "riveting from the space shuttle" thinking there might be some interesting photos of astronauts doing riveting repairs in space. This turned up in the 19th ranked entry: *Friction welding as a rivet replacement technology, patent no. 6779707.*

To see what that patent is all about, I go to the Google patent search site http://www.google.com/patents and type in 6779707 in the information search bar and up pops a 23-page patent. I go there, and look at the drawings ("figures") in the patent. It appears that this sophisticated process for joining aluminum alloy panels such as used in aircraft would not be easily visualized by just looking at conventional rivets. A frictionally heated tool is pulled along to continuously weld two panels together, replacing the conventional rivet process. Inventing this process would require conceiving the welding process as seen by the welding tool as well as understanding many things about aerospace materials and fabrication processes.

Key Points in Using Google and the Clouds

Google can be extremely helpful in helping you find your way in some of the stages of your new creation or invention. Google, however, will not do your abstract thinking for you. Google compares data sets exactly as they appear. It does not do abstractions of the data sets. The human mind, by comparison, has powerful imagination in part because it can do abstract thinking stimulated by the data sets. Abstract thinking and reasoning belong to a set of cognitive skills that are reserved for higher level thinking not yet found in the computer clouds. These abstract thinking skills are required for original and creative thinking.

When you think of a bright idea that you want to research further, you want to do it now. Your neurochemicals are flowing. Your neurons are alertly looking for new connections.

Your brain steps up to an emotional high. While it is there, avoid procrastination! Keep moving while you can see all of the project's interrelationships. While fear of failure is undoubtedly present, on balance you will feel better about yourself if you pursue the emerging development while your mind is in high gear.

While you may not get a new insight out of Google right now, by entering its search cloud with your new project in mind, you have that exciting feeling that you have inexorably launched the development project. If your creative consciousness has been raised, then you are making progress. This should lead to raising your enthusiasm, curiosity, sense of purpose, propensity to ask profound questions, and bold creative thinking.

You should not overlook using Google or other search engines and Web resources to research the early history of your field of interest. This history notably would include the pioneering big thinkers in the field. It can be very revealing to see what they thought about the important issues. The original thinkers in various areas were, by nature, very curious and often expressed themselves in a frank and wondrous manner. While their naïveté in hindsight may seem laughable, on the other hand they often spoke in straightforward terms that paid no attention to offending colleagues or to political correctness. They noticed things that might be worth reexamining.

My Experiments with Internet Design

My first brush with the Internet and creativity came in the late 1980s when I asked my Chinese immigrant engineer his solution to solve a particular aspect of a major product development problem. His answer was unexpected and unsettling: "I will find it on the Internet!" I was shocked and dismayed that its solution might be found somewhere in cyberspace. I did not realize the extent of the information available on the Internet in those early days, and my self-image as the all-knowing inventor was challenged.

My engineer looked like a kicked dog after he sensed my reaction. I thought the Chinese had inscrutable ways of problem solving combining centuries of invention with a lot of training in our universities, but I never expected this mode of problem solving.

We finally developed the product using an innovative key component, and I built a profitable business based on it. This was the plastics gels detector described in chapter three.

Years later, I tried reinventing the product I was discussing with my Chinese engineer. I put in Google search combinations of terms relating to my invention such as gels, particles, polymer stream, seeing transparent particles, fiber optics, process windows and so forth. The searches did turn up some abstruse technical papers in a somewhat related field, infrared analysis, which could give clues to what I was trying to invent. An English product also turned up which incorporated part of my invention but was not on the market when I introduced my product.

I found that I could not "invent" the product in terms of Google presenting me with an "aha!" insight. What Google did do was greatly speed up "what if" searches for engineering and production solutions to building a feasible prototype. This is important because if I could not find an apparent product construction that worked, I might normally abandon the project. The search encouraged me to think that I had a potentially valuable invention and that I should press forward.

I had saved myself hours or days of time of literature search, phone calls, correspondence and meetings. Having established that the basic product concept could probably be converted to a working product, I could pour renewed energy into thinking about everything from the concept to detailed engineering. Also helpful was that the web searches gratuitously turned up some patent applications and issued patents in related areas.

Buoyed by the encouraging results with a Web search engine, I tried using Google to reinvent the wheel. The process was inefficient and strained. Google works better with complex inventions where some law of large numbers allows it to dig out and relate many data. It is not well suited to conceptualizing a few basic relationships like a wheel, hub and axle.

Development success using a search engine is unlikely to happen unless you know the right questions to pose to the search engine. These insights come from experience in many technical fields and probably some trial and error experimentation to define the research approach. I think Thomas Edison would agree.

7

ARTIFICIAL INTELLIGENCE: CAN IT DO OUR INVENTING FOR US?

The non-biological intelligence created in that year (2045) will be one billion times more powerful than all the human intelligence today.
Ray Kurzweil, Futurist

Artificial Intelligence has been used and proposed for a wide variety of projects ranging from designing better scissors to formulating new molecules and drugs. What is AI and will it be a practical assistant?

Arthur C. Clarke in his 1968 science fiction epic film *2001: A Space Odyssey* has his super computer HAL ask the space ship commander: "By the way, do you mind if I ask you a personal question?...I've wondered whether you might be having some second thoughts about the mission." In this case, fiction, written by a scientist, proposes that a computer can think like a person.

According to the Turing Test, which is the standard of comparison for computer intelligence, HAL the computer was most probably as intelligent as a human because the blindfolded observer would not be able to decide which speaker was the computer and which was the human.

Technology futurist Ray Kurzweil asserts that strong artificial intelligence and nanotechnology will be able to create any product, any situation, any environment that we can imagine at will. And this will be before the end of this century. Computer expert and intelligence theorist Jeff Hawkins disagrees. He states, "AI suffers from a fundamental flaw in that it fails to adequately address what intelligence is or what it means to understand something."

First, let us dispel the startling statements and popular movie themes telling us that artificial intelligence will greatly exceed human intelligence in just a few decades. There may be little doubt that this will be the case for applications mostly requiring massive and repetitive computing, but is not so certain for projects requiring significant imagination and creativity. In any case, it is highly unlikely that androids will be running around conquering the world.

AI computers can access very large databases. They can be used in detailed multidimensional design. They can manage vast projects. There is talk of computer-like nanorobots that can circulate around in your body. There are even computer programs to invent new devices. However, as far as I am aware, no computer independently came up with the general theory of relativity.

Since the explosion of electronics beginning roughly during WW II, there has been ever increasing interest in thinking machines. AI started with conventional electronic circuits wired to achieve an intelligent task. These circuits were built on "if...then..." logic. For example, "If the temperature is below 72 degrees, then turn on the heating; otherwise, leave the furnace off." In other words, the traditional home thermostat is a simple AI device. Simple electronic control circuits are the key to many successful industrial machines and consumer appliances today.

Increasingly, tiny single purpose computers called microprocessors or embedded systems have replaced electronic logic circuits. These can introduce more intelligence into the control system. Now we take for granted the availability of home controls that not only control temperature, but also manage the security system, lock and unlock doors, water the lawn, and remind you to walk the dog. Current model automobiles all depend on microprocessors for their operation and maintenance.

The penetration of simple AI is pervasive. Just as there are many times more insects than mammals, there are many more simple artificial intelligence devices than there are complex thinking computers. Cars, airplanes, boats, and space vehicles utilize countless simple AI control devices. These devices have an IQ of less than an ant's, but the technology has been perfected to a high degree of confidence, reliability and low cost.

Major opportunities to be creative in developing simple AI inventions today are in the nanotechnology arena. This is where the mechanical parts and circuits have been reduced in size to almost molecular scale. Applications range from smart phone controls to nanorobots cast adrift in the body to do good things such as attacking germs and tumors.

South Korean scientists are developing a treatment for cancer that is more efficient than chemotherapy. Microscopic robots carry the drugs. They are carried in modified salmonella bacteria, which are drawn to cancer cells by chemical attraction. They attack only the tumors, and the patient does not have side effects like losing hair whose cells normally would also be attacked.

Complex AI

If artificial intelligence could solve more complex problems and do large-scale useful jobs, this would be a payoff for humanity. These challenges could range from diagnosing diseases to managing manufacturing plants. I call this complex AI because complex computers and sophisticated sensors are involved. There will be ever more immense amounts of data, primarily stored in the computer clouds, and there will be software that sharpens its intelligence through continual learning. This can all be part of the Knowosphere.

The wakeup call came when IBM's Deep Blue supercomputer defeated Grandmaster Gary Kasparov in chess in 1997. Deep Blue's per circuit processing speeds are many fold faster than Kasparov's neuron processing speeds. It could examine more than 200 million chess moves per second!

The public really became aware of big AI when IBM's ultra super AI computer, Watson, beat two former winners of the television quiz show *Jeopardy!* Watson received the first prize of $1 million. The system is built around massively parallel processors. It uses IBM's *DeepQA* technology to generate hypotheses, gather massive evidence, and continue analyzing the data until it can propose a solution. The system can process 500 gigabytes, the equivalent of a million books, per second. This program generally does not compute exact answers; it searches for the highest probability answer or solution. Watson was not designed to be creative, assuming that was possible.

Watson used encyclopedias, dictionaries, news sources, literary works and much more as sources. Watson would have to respond to the quiz questions in a few seconds, which, at the time, was thought impossible to do, but the machine clearly succeeded. IBM's first commercial applications of Watson have been mainly in healthcare. With its natural language communications with users, hypothesis generation, and evidence-based learning, it is a natural for use by medical professionals.

Another IBM supercomputer, Blue Gene, is analyzing the brain's structure and operation. It is analyzing its own creator! In 2005, the project was initiated at the Brain Mind Institute of the Ecole Polytechnique Fédérale de Lausanne in Switzerland. The computer can do up to 22.8 *trillion* operations per second. Each of its microprocessor chips simulates a single neuron, which is the basic connectivity element in the brain. Each human brain has billions of them, which makes it too big even for Blue Gene to simulate, so Henry Markram, the project's director, decided to simulate a rat's brain.

Actual simulations were run in mid-2008. The computer simulation produced behavior like a real neuron. Various cells did what they were supposed to do and in the proper sequences. The microprocessors were hooking themselves together and evolving into a cognitive mechanism similar to the core part of the mind. The research team was very encouraged and felt that Blue Brain appeared to demonstrate that their neurological model of the neocortex was correct.

However, scaling the model up to human brain size, where trillions of synapses are involved, would require about 500 petabytes of data (equivalent to 200 times Google's current total system wide computing capacity) and megawatts of electrical supply. What this means, at least for the present, is that current simulations of the human brain have come to a capacity limitation even for a super computer.

IBM's current research is to design systems that can learn from and interact with people. As Guruduth S. Banavar, director of IBM's cognitive computing research, told the *New York Times*, "The result should be way better than either a human or a computer system can do alone."

Brainlike Computers

Another approach in artificial intelligence is to avoid the use of precisely programmed digital computers and instead make an electronic computer that emulates the neurons and synapses of the brain. The strength of connections between neurons represents the relative strength from associating ideas, places, words, and the like. These connections can be strengthened or weakened with experience. This form of computing, sometimes called associative inference or thinking in analogies, is inexact compared to digital computing in standard (von Neumann architecture) computers but it is often much faster, uses much less power, can analyze with very incomplete data, and works despite some damage and noise in its structure.

The principles have been known since the artificial neuron experiments with simple electronics after World War II. Brainlike computers, often called cognitive computers, have only been built recently, however. Stanford University, California Institute of Technology, IBM, and Qualcomm are among the leaders with cognitive computer projects. These computers and others of their kind probably will be the basis of the most general application AI, which will be equivalent to the overall intelligence of the large mammals and humans.

In August 2014, IBM announced their brain-inspired computer chip. Called "TrueNorth," this chip has a million neurons—about as many as a honeybee. It carries out 46 billion synaptic operations per second in its 256 million programmable synapses and can be powered by a hearing aid battery. Its 5.4 billion transistors fit into a space the size of a postage stamp.

An important application for TrueNorth is video pattern recognition. It should be able to recognize people and objects in a scene. It should also do better speech recognition than the present systems, and consequently should find an immediate home in smartphones.

IBM has a goal of supplying devices for a 100 trillion synapse supercomputer—about the synapse count in the human brain—and at that point we will be well on our way to making HAL in *2001: A Space Odyssey* or the female computer person in the movie *Her.* Its energy consumption would be considerably less than a digital computer of similar capability, but much more than the human brain (20-40 watts). The power consumption will be an area for further improvement of the technology.

Among the variety of applications proposed for the IBM TrueNorth chip recently are:

- Provide sensory inputs to guide a blind person like a seeing eye dog would.
- Automate the surveillance input in real time for military and commercial drones.
- Serve as the core of a laboratory instrument to help scientists simulate brain functions.

Inventing with AI

Inventing by use of an AI program has succeeded in a number of cases where the product to be invented involves one discipline such as mechanics or electronics and does not involve too many component parts. This emergent field has taken on the terms genetic algorithms or evolutionary algorithms because the design approach is evolutionary, seeking optimal success as various mutations are introduced.

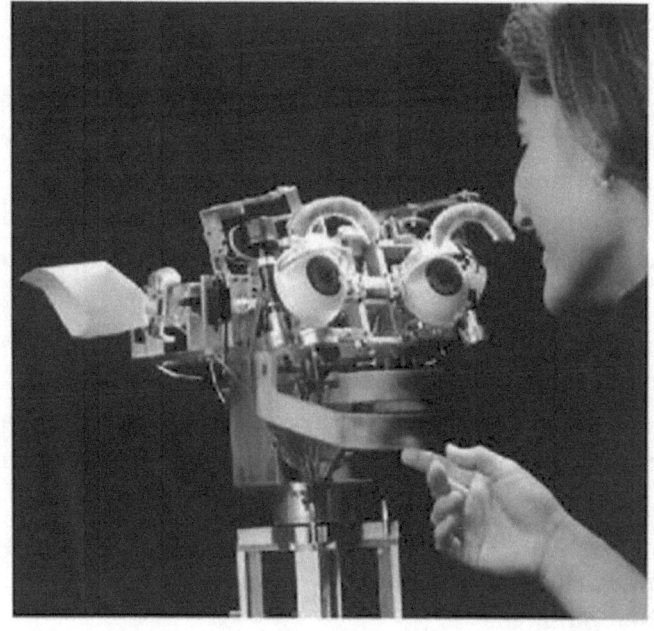

©MIT 2010

The human-robot partnership.

Can we be a powerful creative team?

While genetic or evolutionary algorithms are usually best suited for the design of simple devices where one discipline is involved such as an electronic circuit of a few components, or the shape design for scissors, their use has been reported for the design on dams, bridges, gyroscopes and wind turbines.

Ed Ramsden, writing on "evolutionary computation" to develop optimum signal processing circuits for sensors, calls this "Just add parts and shake." He wrote that the following steps are involved in designing an optimum circuit such as a filter (A filter is a circuit to separate a desired electronic signal such as music from the unwanted signal or background noise) using just computer simulation:

1 – Define the basic circuit in terms of the how the components are connected to each other and the component values. It will be optimized by trying out variations of its design.

2 – Define a criterion of optimum circuit performance. Maximization of results based on this criterion indicates when to stop trying design variations.

3 – Start with a circuit design using electronic component values chosen arbitrarily.

4 – Simulate operation of the circuit, change component values and their placement in the circuit, and record circuit performance at each step.

5 – Rank circuit performance in terms of design changes. Eliminate all but the front-runners. Select the best circuit from the final candidates.

You could do the same steps in perfecting a chocolate chip cookie recipe—baking cookies many times and changing ingredient proportions, baking temperatures and times, etc., until the cookies were delicious beyond description. This process would be very difficult to adapt to artificial intelligence because it would be difficult if not almost impossible to design a machine-readable measurement for the "most delicious cookie."

Designing robots is an application of this approach, and there are reports about self-improving robots. Some robots design themselves—especially their control circuitry and programming—based on their trial-and-error results.

For example, you could start with a collection of shafts, wheels, sensors, structural kit parts and so forth and build a robot as a starting genotype as described for circuit design above. Then you could observe the robot interacting with its typical environment. You then could change to its design, for example, reducing its bumping into things or experimental random changes in the design to see what the change in performance will be. Developers of self-improving robots claim that they produce sophisticated robots that work as well as the ones designed by humans. Many examples and reports can be found by an Internet search.

It should be noted that these applications avoid having to deal with mechanical engineering, materials, fittings, etc., whose selection and optimization is the hard work of many inventions. This is consistent with my own experience with three patented inventions that involve chemical engineering, mechanical engineering and electronics all blended together. The Patent Office wants to see evidence that your creative insight can be reduced to practice.

A fundamental weakness so far of the AI invention process using genetic or self-improving algorithms is that it only works within a closed system. The program is given a list of components to interconnect and optimize, but so far, it does not allow inquiry into other methods or improvements to better achieve the invention's goals. It does not allow intellectual inquiry into all the resources, methods, and prior art found in the computer clouds.

What Went Wrong with AI and Where Next

Human thinking is flexible and accommodates ambiguity, so strong artificial intelligence must include probability in its conception. It must look for patterns in masses of data. This development approach leads to structuring artificial intelligence based on biological models, like the human brain. We have been hobbled by our attempt to create intelligent computers, which operate by traditional step-by-step analytical logic.

Artificial intelligence presently works for smaller problems with more achievable goals. Successful applications include medical diagnostics, stock market trading, optimal vehicle control, mineral prospecting and home appliances. At best, there are AI controlled machines that can automatically do limited tasks such as vacuuming a floor or guiding a rocket.

The AI computers and systems up to this point have generally been judged to be disappointing in terms of the concept of devices that would mimic human thought and intelligence. It became clear that artificial intelligence programs were limited in scope because they incorporated step-by-step computational algorithms based on reductionist conceptions of how to solve problems. The models could not reach out and incorporate new variables or data that the model itself thought might be useful, let alone do anything abstract.

Further, the models usually did not assume that random changes in the situation from time to time could change the behavior of the system being modeled. For example, if we have a model of sunlight affecting an insect, we must assume that on some days there is no sunlight and on some days the insect does apparently random acts of non-obedience to the model.

In other words, we will have to deal with the real world where almost any situation of interest for resolution by very intelligent machines will have hundreds or thousands of variables. These variables do not have static relationships. Their interrelationships are constantly changing with time.

It is like looking at the morning television newscast showing a live video of traffic seen from a helicopter. While the flow of cars is the same from day to day at commuter hour, we observe that the flow volumes on given roads change depending on accidents, weather, special public events, etc. To model this traffic flow so that it can flow more efficiently when controlled by an AI computer gets more and more complicated as we peel back the layers and find ever more variables. This is also the case when modeling climate change.

The human brain gets around a lot of this complexity problem by storing many patterns for future reference. In the traffic analysis example, the experienced helicopter pilot would have stored thousands of patterns and the associated information about weather and the many other variables. Without doing the time-consuming exact computations that a computer would do, the pilot would project his thoughts based on analogies with historical patterns. This is an example of associative inference where the brain thinks in terms of patterns (not variables) stored associatively (related patterns are stored together) in the neocortex.

So far, though, we really have not confronted thought head-on, whether it is done in a brain or a computer. David Gelernter, professor of computer science at Yale University, writes:

> AI has no comprehensive view of thought: it tends to ignore some thought modes (such as free association and dreaming), is uncertain how to integrate emotion and thought, and has made strikingly little progress in understanding analogies—which seem to underlie creativity.

Gelernter goes on to say:

> Computers don't know or care what instructions they are executing. Switching applications changes the output, but those changes only have meanings to humans. *Consciousness*, however, doesn't depend on how anyone else interprets your actions; it depends on what *you yourself* are aware of.

One of Gelernter's constructs is the *cognitive continuum* where all information you receive and remember is cognitively connected to all other information you receive and store. It connects such reasoning as analytical thought, analogical thought, free association and creativity. These thought processes all are part of mental focus or concentration. Without this cognitive continuum, he says that AI has no comprehensive view of thought. Gelernter's comments offer a sobering perspective of how far AI has to go.

PART 2

BIG DATA AND NEW SOLUTIONS

8

TEAMS, COLLECTIVES, AND THE CLOUDS

The open society, the unrestricted access to knowledge, the unplanned and uninhibited association of men for its furtherance—these are what may make a vast, complex, ever growing, ever changing, ever more specialized and expert technological world, nevertheless a world of human community.

J. Robert Oppenheimer

As I was leaving childhood, it was the 1950s and 1960s. That was a carefree time bridging the national self-confidence after World War II with the hope coming out of the labs, big cars with tail fins, "Atoms for Peace," miracle drugs, and electronics. The world was not ready to think about the big ecological picture except for some strange voices like Rachael Carson who wrote stirringly about impending ecological crises.

I recall the excitement when computers became useful in the early '60s. As an engineering student, I learned to program those huge machines full of glowing tubes. A coffee pot was kept warm on an equipment rack. Replacing burned out tubes was as routine as sweeping the floor. Transistors were just being invented and circuit chips were a figment of someone's imagination. Data was fed into and out of the machines by punched paper tape and later by the familiar IBM Cards with the rows of rectangular holes. The most advanced machines, accessible only by pedigreed top researchers, had less computer power than today's personal computers.

At MIT, I was an assistant to Professor Franco Modigliani who later received a Nobel Prize for his Life Cycle Theory of Savings. This theory maintains that the level of savings depends on the age of consumers rather than on the level of family income. The very young and the aged draw down on family savings while the working age members produce the savings.

He would call me anytime, day or night, "Larry, it's time to run some more data!" This always seemed to happen on Saturday night, and my date would help sort punch cards into decks to feed the hungry computer. The next day, Sunday, there would be printouts to review with the professor. Modigliani would review the correlations, modify his model, and I would head for the patient computer to do another run. The world at that time was ready for econometric-based economic theory but not major complexity.

A few years later, I graduated and took my first real job with a Cambridge, Massachusetts think tank called Arthur D. Little, Inc. One of my projects was working with an antisubmarine warfare team. We designed long-range sonar for tracking Russian submarines, and we devised mathematical scenarios for predicting how the Russian submariners would try to traverse the oceans to the US. "He thinks that we don't know that he knows a special route free of sonar…" The thought gaming simulations were entered by a clackety-clack teletype to a time-shared sort-of-super computer far away. A laptop would probably better handle this analysis today.

I had found myself in the early stages of the analysis of large scale, complex problems that today has engendered its own scientific discipline called big data. Solving these problems has lead to big team research.

Creative Teams Led by a Charismatic Technical Leader

World War II produced a host of challenges and eventually products that involved complexity on a scale unimaginable in the inventive times of the nineteenth century. The atomic bomb immediately comes to mind, but there were plenty of other big challenges. Important new technologies included antibiotics, plastics, heavy-duty aircraft, rockets, and radar. While one man or woman could have done the initial idea or discovery, the further development of the project would quickly lead to the formation of a team.

The team became known as "The Lab"—a convenient term for an affiliation of engineers and scientists collaborating to solve a specific problem. MIT pioneered this with the Radiation Lab, which was responsible for much of the radar development during the World War II. As an electrical engineer, I found their lab notes published in the "Radiation Lab Series" still useful, for example, in the derivation of optimum filters.

Another MIT mission-oriented lab was the Instrumentation Lab, which developed servo systems and inertial guidance systems so critical for the important national efforts of the cold war ranging from guided missiles to lunar probes. A national project of singular importance and impact was the Manhattan Project to develop the atomic bomb at Los Alamos National Labs in New Mexico.

A key part of the lab organization was having a collection of people working on the problem, starting with a charismatic scientific leader. He or she would have a prestigious board that could include a university president, scientists, industrialists, academics and military representatives. Staffing was often ad-hoc with scientists and engineers grabbed from almost any-where. Shirtsleeves experimentation always was important. J. Robert Oppenheimer famously fulfilled that role by leading his team that developed the atomic bomb.

This same pattern was in the submarine-tracking program I mentioned earlier. Our designs were successful, and the Navy deployed a worldwide system to detect, identify and track enemy submarines. The project, code named SOSUS, was first publicly mentioned in the book and movie *Red October.* Like the World War II development projects, we had a scientist leader, an eclectic collection of support engineers and scientists, and a high ferment creative environment.

Anyone, however lowly, could volunteer an idea. People would dash off tennis courts to scribble down concepts. The really important classified stuff was not determined by the ominous degree-of-secrecy labels on filing cabinets; it was determined by an informal almost secret society of intellec-tual buddies. As the project became institutionalized, with a management bureaucracy taking over, the old gang including myself drifted off to seek new challenges.

It seemed that the bureaucracy evolved from fostering inno-vative solutions for research goals to preserving the status quo and raising more financial support. While researchers under-stand the need for stability and funding, they also seek leaders who are creative or at least understand the creative process.

Innovation Through Informal Encounters

We had a study group researching how to manage innovation most effectively. An interesting finding was that the prob-ability of coming up with a major breakthrough was directly proportional to the proximity among the team's scientists, engineers and technicians. Everyone located on the same floor of one building lead to the best results. Chances of success declined as the project occupied two floors of a building, two buildings even less, and least of all, two sites a day's or more travel apart. The conclusion was that innovation is fostered by informal communications, and these are encouraged by many informal "what if?" and "let me show you something" discussions.

The restraints of not working on the same floor or in the same building may be relieved by Wikis (common interest Internet discussion groups). Web-based development collaboratives, mobile phone hookups and the like are all attempts to foster group creativity and informal communications.

The lab development organization carried over to US and other countries' corporate development projects largely for non-defense products. IBM, Xerox, and Bell Labs were among the industrial companies that produced troves of valuable patents that were critical to industrial growth and consumer well-being in the 1960-2000 period.

One indicator of the continuing trend towards large group research is the ever larger number of names listed as authors on research papers. In the old days, 3-5 might have been considered a typical research team. Nowadays it is not uncommon to see over a dozen names and some reports have listed over 100 or even 1,000 authors.

The major development opportunities and priorities now are molecular biology, computer and software systems, energy, climate change, and ecology. All of those areas typically require research efforts with at least dozens if not hundreds or thousands of scientists and engineers per project.

The ideas these teams work on can start with seeds planted by inventors. These turn into green shoots if they are encouraged by the team. If success continues, management and investors take control, hoping for a big harvest.

The Emergence of Connected Intelligence via the Web

When it seemed that much important technology development would be confined to research parks in known creative places like the Boston area, the Bay Area, and the Research Triangle, the emergence of the Internet made physical location a much more flexible option for many participants. A leading scientist might work out of his home office in the Great Smokey Mountains and teleconference daily to his labs in San Diego and Cambridge.

It is important to note that all communications media can be used. It is no longer just emails, although they are still important. Video conferencing is simple and low cost. Laboratory experiments can be read and managed from anywhere. All sorts of messaging formats are available for Internet communications.

Perhaps most important, the remote scientist can access all the world's libraries through the computer clouds using Google and other search engines. He or she should also be able to access their employer's library and wiki through a secure broadband data connection.

All of this began inauspiciously when Larry Page and Sergey Brin met as students at Stanford in 1995. Brin was assigned to show Page around campus. In 1996, these two computer science graduate students collaborated on a search engine called *BackRub* run on Stanford computers. The following year they changed the name to *Google* based on the mathematical term googol meaning 1 followed by 100 zeros. This inspired term represented their goal to organize an almost infinite amount of information on the Web.

In 1998, they incorporated their venture, which was operating out of a garage in Menlo Park, California with $100,000 of a friend's investment. The number of URLs (websites) that were indexed by Google grew exponentially from about a billion in 2001 to about a trillion in 2008.

Early in 2008 Google came out with its Cloud Strategy. The concept is to deliver to students, researchers, and entrepreneurs all the power of Google computing using a cloud of Google and other computers. Their cloud is about a million networked small computers like ordinary PCs. They are connected to the Internet and answer billions of requests in fractions of a second.

The energy consumption of all these computers is so significant that Google has researched and designed a new computer basing approach. The computers will be put in a ship at sea where the ocean currents will provide energy for generators and water for cooling. A patent was issued to Google for this interesting idea.

Massive (essentially infinite) computing power becomes an essential resource in this emerging era of connected intelligence and the collectivization of information. Software for accessing and working with the information is also of critical importance. Google corrects spelling and translates languages with a program that fits on one sheet of paper. This is important because that same program must be used millions of times a day delivering information to users in a fraction of a second.

The verb "to Google" has entered the common vernacular as the preferred term to do a Web search. In the rest of this book, I will often refer to Google, and about using Google, as generic terms to this search methodology and database when it is to be understood that various other search engines could serve as well. However it is accessed, the World Wide Web has put most of the world's published knowledge at your fingertips.

9

FINDING SOLUTIONS IN OCEANS OF DATA

In the era of big data, more isn't just more. More is different.
Chris Anderson, *Wired* Magazine

"There is such a thing as everything" is what I overheard while passing a boy of about seven playing with some friends on a hiking trail. They were looking at flowers, trees, and some butterflies, and were awed about the whole ever-changing scene. For a youngster who has not confronted the practical limitations of computing, this seemed like a perfectly reasonable thing to say. I think Leonardo da Vinci would have happily endorsed the boy's point of view.

However, for us, the older and wiser laborers in the vineyards of knowledge, the idea of looking at something based on all we know about the whole world in order to understand and model a part of it seems at best to be a kid's fantasy. It is of little practical value over existing modeling techniques. Even super computers have not reached the point where we can use them to model the whole universe.

However, with new information technology, the holistic all-encompassing view has several positive points in its favor:

1 – *The almost infinite search engine database.* The enormous, ever growing, and instantly accessible knowledge base accessed through Google, Microsoft, Yahoo, Amazon and others (what I call the Knowosphere) is so fast and comprehensive that it is a new resource and not just an electronic library. It has information that is required to refocus and refine the individual's thinking and thereby multiplying their mental powers manifold times. Sitting all by themselves, the computer clouds will not invent anything, and sitting all alone, the individual will not realize their full potential as an inventor. However, the combination of both can produce an invention of immense scope and importance.

2 – *Relationships derived from massive sets of data.* Peter Norvig, Director of Research at Google, implemented and popularized an interesting and important discovery. He confirmed that using a corpus of millions of words indicating the probability of their co-occurrences (the probability of any two, three, four or five given words occurring together in a sentence), he could predict the correct word from a misspelled word with a prediction accuracy considerably better than chance. This program is part of the familiar Google spell checker and fits on one page. Norvig also implemented the program that translates the texts from one language to another using the co-occurrence corpuses for words of both languages.

In mathematical terms we can say that Norvig's revelation was that the data teaches us the algorithm if there is deep enough inherent structure and if there are indeed relationships between sets of data. In general, if we work with massive data sets we can often use different and relatively simple computational algorithms whose computational results are not exact answers but are good enough.

The traditional scientific approach has been to use exact computational equations or algorithms and assume that only a fixed number of well-defined variables are involved. This is the classical analytical approach and is often referred to as reductionist. It is the general algorithmic approach used in many classic artificial intelligence programs.

Machine Learning: The Computer Develops Its Own Model from the Data

Analysis using multi-million cell data sets is ever more important because many of the world's most important research and development challenges are problems involving dozens to thousands of variables and thousands to millions of data points. Analysis of many data sets produces likely candidates for variables in a system model. The model building uses inductive reasoning where relationships suggest themselves based on observations that statistically seem likely. These problems are likely to be solved empirically where the computer searches for a best-fit model in a process called "machine learning."

An interesting example of machine learning is its use to analyze complex biological systems. In the development of a model called MEDUSA to determine how an organism's genes behave under various conditions, Chris Wiggins at Columbia University is using machine learning to see which pairings of sets of DNA most influence the activity of the yeast's 6,200 genes. For example, when pairings are changed in the model, effects are observed if RNA production is increased or decreased. Using a statistical approach, the behavior of the organism's gene regulatory networks can be determined. Of course, any given statistical prediction is not 100% certain, but certainty is only relative in biology where no two organisms, though theoretically identical, are exactly identical. In any case, the emerging model for yeast has been helpful to biologists in determining the collection of proteins that relate to a cell's features. Wiggins points out that "machine learning lets the data decide what's worth looking at."

Storing of Billions of Data Sets by the Brain

The human brain, on the other hand, rather than storing a huge pile of impersonal data and then finding patterns in it, stores all input information flow from the outset as patterns. Everything we see, hear, smell, and read is received somewhere in our brain and stored as patterns or additions to existing patterns. The information is connected by activated synapses to previously stored information as interconnected associations of similar percepts (color, person, city, time, etc.) as discussed in chapter five. Information storage can be in the trillions of bits. By the time you are six years old you are considered independently intelligent (you must start grade school then), and one reason is that your number of stored precepts, concepts, events, etc. must total in the millions. By the time you reach old age, those remembrances probably total in the billions or trillions. This is important to have some statistical certainty in your thinking process.

A great feature of the human brain is that it does not require complete information about a given pattern to use that pattern in its thinking. It "fills in" the missing information to be able to continue thinking without delay. The brain remembers the important information as a priority over remembering all the details.

Human intelligence really is not meaningful as a static model. It reveals itself as a process where new information flowing in is integrated into a greater whole. The manager (you) is aware of what the on-going assembly looks like and can use that as a database for referencing new incoming data. Changes noted in the pattern comparisons give the brain a basis for predicting future changes in the patterns.

Comparing the Computer with the Brain

As we review what has been said so far in this chapter about massive computer databases, machine learning, and pattern learning of the human brain, I propose that a fundamental difference persists: we probably cannot build and program a digital computer that totally emulates the mind nor can we force the mind to do all the things the computer can do. At first glance, the emulation of one by the other looks easy, but there are irreconcilable differences:

1 - *Data Formats.* The mind can do digital computations like a very simple digital computer (1 + 1 = 2) but it is by far more suited for verbal, aural and visual information processing. These are precisely the data formats that computers struggle to use. I myself have engineered computer systems to notice and analyze changes in a video scene. It can be done, but the electronics and software are daunting indeed compared to what even a mouse (the animal) can do in this context! On the other hand, the human brain cannot compute the square root of two to the high precision and in the short time achievable by even a $10 pocket calculator, let alone do the simplest tasks achievable by any laptop computer.

2 - *Open sourcing and indefinite expandability of information.* The mind is programmed to store information for all kinds of eventualities. This is a major driver for "natural curiosity." Even dogs, for example, have this as can be seen when they investigate any new space for food, doors, friends, threats and places to sleep. Humans have the higher calling of the continuous expansion of the mind's library of knowledge. Computers, on the other hand, have no survival-based instinct for continually expanding their information wherever such information ingestion may lead. They simply absorb whatever information is supplied to them if it fits their data formats and storage capacities.

3 - *Processing speeds and capacities.* Traditionally, in artificial intelligence discussions, the idea of computers replicating the human brain and indeed exceeding its capacities was brought up and agreed upon as a theoretical possibility but not likely in the foreseeable future. Ray Kurzweil, inventor and artificial intelligence optimist, created a stir in his book *The Singularity is Near* by saying that by 2030 that computer capacity will be achieved which is equal to the human brain. The "Singularity" he refers to is when computers will attain a profound expansion of our intelligence, and this "profound and disruptive transformation in human capability" will happen in 2045. However, Kurzweil reassures us that "Despite the clear predominance of nonbiological intelligence in the mid-2040s, ours will still be a human civilization. We will transcend biology but not our humanity."

Limitations of Computers

There are at least two problems with this vision of the future. One is whether computer capacities will continue to climb rapidly, and the other is if the computers of the future will be able to incorporate such apparently essential human thought attributes as consciousness.

Kurzweil bases his technology projections on a number of data plots that compound as straight lines on a logarithmic scale. The best known of these is Moore's Law named after Gordon Moore, a famous inventor of the early integrated circuits, who later was chairman of Intel. Moore observed that the number of transistors, a basic element in computer logic circuits, in an integrated circuit doubles about every two years. Combining this with several other technical considerations, Kurzweil arrives at a projection that computer price-performance doubles every year.

Recent publications by computer scientists and engineers cast some doubt on this prognostication by pointing out that even if computer designs incorporate quantum spin technology—thereby reducing what was the transistor to an atom—that other limitations come to the fore perhaps most importantly heat dissipation. Some large, fast computers today are water-cooled and some emerging designs will be cryogenically cooled.

As more and faster computation cycles are crammed into shrinking volumes, something has to give way and that appears to be heat dissipation. I started my electronics career at the end of the vacuum tube era, and at that time saw transistors take over, and later integrated circuits—millions of transistors on a chip—so I intuitively feel that within a few decades there will be technical breakthroughs which enable computers with human brain capacity.

The bigger problem standing in the way of Kurzweil's vision is the human brain algorithm—how it thinks and its on-going process of deciding what to think and what to do. As described in the preceding two chapters, the brain combines all its stored and incoming sensed patterns in as many ways and combinations as possible. It constantly is alert to new information to improve its understanding of its environment or of a goal it has set for itself.

Association and abstract thinking become important. Through consciousness, the brain reviews all relevant information in a dialog with itself and then moves forward with the next course of action. This can be done using artificial intelligence to avoid the use of a precisely programmed digital computer and instead use an electronic computer that emulates the neurons and synapses of the brain. The computing time is orders of magnitude faster than a digital computer working on an associative inference task. This is discussed in more detail in "Brainlike Computers" in chapter seven.

Search Engines

What will be available to anyone with any kind of personal computer are the ever-growing, practically infinite, computer clouds of information accessed by search engines. If access to all kinds of publicly accessible information is required to develop a great idea, this capability already exists in the computer clouds. The iterative interplay between the creative human thinker and the computer clouds resource for major problem solving should be the equal of Kurzweil's super mega computer for the foreseeable future.

One such productive information interplay in progress now is at Google where they are trying to make us more productive as thinkers. One of their founders, Larry Page, told a convention, "The ultimate search engine is something as smart as people—or smarter." At a more recent convention, he proclaimed that Google is "really trying to build artificial intelligence and to do it on a large scale."

Undoubtedly, we will greatly benefit from the information resources available to us through the computer clouds. Inventors must use resources beyond the traditional workshop and laboratory. Major technical discoveries and developments are now being done as laboratory projects by science and engineering teams. Now major developments in such complex areas as alternate energy sources, drug development, and large aircraft design require not only the classical talents of inventors, scientists and engineers, but also the combined resources of libraries, laboratories, field studies and all kinds of computer and software resources. The focused thinking of the creative human brain and the human guiding of Internet search is potentially more intellectually productive than either developmental method alone.

Examples from Drug Research and Development

The first practical example I heard about this new medical treatment was on National Public Radio, May 8, 2008. It is simple but illustrative of the process. A family's doctor, Hal Dietz at Johns Hopkins University who specializes in Marfan syndrome, was looking for new treatments that would not require surgery. Marfan syndrome is a genetic disorder of the connective tissue resulting in defective heart valves, aorta and other organs. It affects about .02% of the US population so it has not received the drug discovery research efforts dedicated to major diseases.

Dr. Dietz knew that a protein called TGF-beta seemed to be active in people with Marfan syndrome. The aorta in mice with TGF-beta blocked developed normally. Where could he find a TGF beta-blocking drug? Typing "TGF beta-blocking drug" into a Google search turned up references to a drug called

Losartan used as a blood pressure medication but also known to be an effective TGF-beta blocker. Losartan had been used to treat tens of millions of patients with high blood pressure and is noted to be extremely safe and well tolerated.

Because Losartan already is in pharmacies and has been tested for safety in humans, Dr. Dietz can write prescriptions for it. While Dr. Dietz's story represents a very elementary example of discovery using Google search, it does represent an approach which can be used on a much larger scale for inventive research.

One example of a new technique for drug discovery that could be used on a very large scale for drug discovery is "systems biology." The databases are so large and rapidly changing that some data storage and search mechanism like Google will have to be used. Systems biology is a holistic approach to incorporate all elements of a biosystem—typically a cell—and to model the interaction of each element with all other elements. The disciplines include biology, chemistry, physics, math and computer science; and the biological elements include genes, proteins and enzymes.

"The models that will come out of biological systems will be the most complicated models that have ever been built" states Dr. Peter Sorger of MIT's Computational and Systems Biology Initiative. "It's going to be the development of fundamentally new paradigms and approaches to modeling that we've never seen before."

The website of the Harvard Medical School Fontana Laboratory, which is focused on systems biology, notes:

Empirical knowledge about protein-protein interaction is rapidly evolving, while being scattered across different research communities. Models and the facts they rest upon must therefore become self-documenting. They need to be embedded in an "operating environment" designed to help biologists cope with incomplete, inconsistent, and continually changing information. Modeling in biology is also a process for inventorizing knowledge.

One project for centralizing knowledge on cellular biochemistry and the modeling networks of cell signaling is called Cellucidate. Professor Walter Fontana, Professor at Harvard Medical School's Department of Systems Biology and at the Santa Fe Institute, calls this "A Facebook for researchers that deal with proteins." He adds, "People are very skeptical about modeling. They say we cannot model because we do not know everything yet. But this is precisely why we need to model."

To get the feeling for the scope of a drug-screening project, consider that it would not be unusual to put a million different molecules through a screening to detect a target response. Systems biology seeks to replace this random testing approach with a more informed approach to discovery and development.

The computer clouds that could incorporate the ever-growing database for drug discovery using systems biology can be "public" or "private" or both combined. Drug giant Pfizer has created a collaborative Internet website operating a combination of resource library, chat room, blog and bulletin board—a wiki—that combines a general map of research going on throughout the company with hypertext links to detailed research. By 2007, it was receiving 12,000 hits per month by 13,000 individual users from all Pfizer operations worldwide. Its community calls it the Pfizerpedia. The Pfizer researchers through Google and other search engines can access most of the world's "public" knowledge. For any particular researcher, results from Google and Pfizerpedia searches can be combined in a dynamic database in their own computer or workstation.

Target-based drug discovery using genome research has been the garden path for many large research groups. The goal is to block or overcome illnesses caused by defective genes. Unfortunately, the number of drugs developed this way has fallen every year.

Important drugs are still developed by old-fashioned inventing. Dan Hurley reported in the *New York Times* that a chemical engineer named Todd Zion developed a new form of insulin. This would be self-regulating. His modified insulin molecule has a sugar attached. With this reactant, the insulin molecule can react with the cell's insulin receptors as required to optimize the sugar levels in the body. Constant insulin testing by the user no longer will be required.

In 2010, Zion sold his company, SmartCells, to Merck. The drug giant estimates the market for his developments to exceed a half billion dollars. Nevertheless, the testing and approval process will drag out market introduction to 2021 at the earliest.

Aircraft Design

Boeing's aircraft products represent a different but no smaller challenge. In this case, for very complex, highly optimized large aircraft, the challenge is not to make sense of millions of pieces of data, as in the search for a drug solution; the challenge is to design thousands of parts or subsystems that will integrate as an optimum total system. Boeing is home to 165,000 employees so designing an aircraft is not like the movies that often show an entrepreneur, a pilot, an engineer and some mechanics in a drafty hanger producing drawings in weeks and a plane in a year.

The basic airliner of our times—the 737—has 367,000 parts, not counting bolts, rivets and other fasteners. The 747 whose economy section is 150 feet (45 meters) long or about the length of the Wright brothers' first flight at Kitty Hawk, NC, has about 10 times the number of parts as the 737. An engineer designing a given part must consider its effect—such as weight, resistance to stress, etc.,—in relation to all its neighbors. The sum total of all parts characteristics, such as weight, must satisfy the design goals for the whole aircraft.

Boeing has moved from paper drawings to computer aided design (CAD) drawings, which are stored as a master set in one computer and instantly available to Boeing people worldwide in what could be called the Boeing cloud. The computer software program, called CATIA, allows three-dimensional graphics for all parts to assure utmost precision and fit. From the very beginning of the design process, each designer and engineering team must react with the Boeing cloud as well as the public Google cloud to deal with the design goals effectively and economically.

The point of all the foregoing is to make the design of something as complex as a new drug or new aircraft as simple and meaningful as, say, designing a hammer. The designer or inventor cannot afford to overlook the heaps of information that is available for any aspect of the project. This survey of the clouds can energize the inventor so that he or she can completely project their mind into the appropriate invention space. They can force themselves into a unique intellectual position where the solution will occur to them and quite possibly no one else. Once found and accepted as feasible, the solution can be optimized and crafted into a practical design, again using the resources of public and private data clouds.

I should confess that, although I try to be a scientist, I am by training an engineer. We engineers are looked upon suspiciously by scientists as grasping at solutions without regard to whether there is solid science behind them. The success of low-brow empirical language translation programs at Google, or the models developed for machine learning for drug discovery, often can help us get somewhere until we figure out the theory. The process is going back and forth between hypothesizing and proving a model and finding empirical data to confirm or reject the hypothesis. In addition to designing and carrying out experiments, researching the vast resources of the Web may hasten confirmation of the model.

It would be interesting to see how aspirin could be developed from scratch today and how long it would take to approve it for general over-the-counter sales. By sorting through various folk remedies in the fifth century B.C., Hippocrates discovered that the bark of the willow tree treated fevers and pain. He had empirically discovered salicylates without a pain chemistry model. In 1897 Felix Hoffmann, a chemist working for Bayer in Germany, devised a milder formulation, acetylsalicylic acid, now known as aspirin, which he used to treat his father's rheumatoid arthritis. It was not until 1971 that a British Pharmacologist, John Vane, found that aspirin controls prostaglandins that control body compounds involved with the maladies treatable by aspirin.

10

ORGANIZING AND SEARCHING DATA

I do not fear computers. I fear the lack of them.
Isaac Asimov

Historians have always benchmarked ancient civilizations by their achievements in collecting, cataloging and archiving knowledge. The library in Alexandria, Egypt was once the Wikipedia of the known world. Destroyed about two-thousand years ago, it housed one-half million to a million scrolls. On the other hand, this vast collection was not available to anybody but only the literate few with time to browse; and it certainly was not available on home computers. It was also hard to find what you were looking for.

Fourteen-hundred or so years later Johannes Gutenberg invented movable type and began commercial printing which quickly spread across Europe. It became the vital catalyst which ushered in the Renaissance and scientific revolution. This was somewhat like inventing the personal computer or the smart-phone in that it was an enabling mechanism for information processing at the local level and making it rapidly available everywhere else.

Recently such basic commercial products as Intel chips, Microsoft software and the Google search engine have enabled the modern equivalent of the library at Alexandria to be instantly accessible on the laptop and smartphone of every man, woman and child in much of the world. Proclamations about the death of the book are premature, but the on-line communal encyclopedia, Wikipedia whose contents probably surpass all print encyclopedias combined, has almost wiped out printed encyclopedias. The age of Google has ushered in resultant fear and misunderstanding about the new paradigm, much like Gutenberg must have experienced.

Information Flows among You, Your Collective, and the Computer Clouds

With the Web and computer clouds, the information flow between you and the computer clouds and you and your collective intelligence groups has speeded up by factors of billions. Humankind progressed from counting on fingers, to the abacus, to the slide rule and mechanical calculating machines, to faster and larger computers. For research and design, this increasing of computation speed, volume and accuracy were needed just to keep up with the exponentially increasing volume of reference information.

The earliest model of information and knowledge storage and use probably would fit in the era from pre-history to the early Bronze Age. You and your clan or village did not have libraries or in most cases not even the written word, but you did have a great knowledge source through tradition, oral histories, and all kinds of know-how passed down from generation to generation.

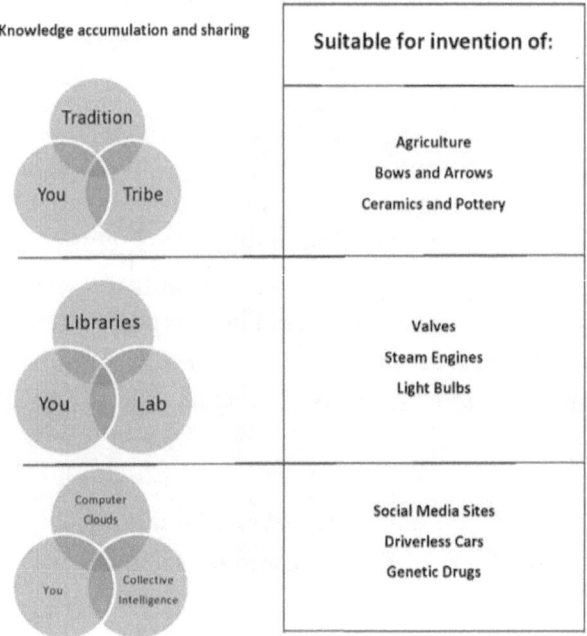

Knowledge accumulation and sharing	Suitable for invention of:
Tradition You / Tribe	Agriculture Bows and Arrows Ceramics and Pottery
Libraries You / Lab	Valves Steam Engines Light Bulbs
Computer Clouds You / Collective Intelligence	Social Media Sites Driverless Cars Genetic Drugs

Throughout history, invention has depended
on knowledge accumulation and sharing.

It is probably impossible to even make a good guess at how much information was guarded by tradition—let alone in a modern metric like bytes—but it was probably a lot more than the typical person of those times carried around in his head. In other words, compared to what he needed to know for daily living, the knowledge in the "tradition cloud" was much more extensive than he would ever need. He knew that chiefs, priests, and clan leaders would have the additional information required for special occasions.

From the Bronze Age to late in the twentieth century, the focus of information and knowledge was the library. The traditional stories and new knowledge of any kind were written down and stored in a central repository. Response to information queries was still rudimentary by Google standards, and it was done through such familiar means as index cards and dedicated librarians.

The 1900's saw the emergence of the laboratory as a local information development and processing resource. It was an important development starting with the major research projects of World War II. At major research institutions, the project "lab" has taken on the meaning as a collective intelligence group. Its resources include a traditional experimental lab but also include a library, meeting areas, and computer facilities—in short, a complete intellectual unit.

Now, in the twenty-first century, your laptop computer, tablet, or smartphone can be the port to all of the world's knowledge. You can locate yourself wherever you feel most creative. Anywhere you can log in to the various computer data resources of interest, you are an active part of the world intellectual community. With petabytes (quadrillions of information chunks) of information at your disposal in fractions of a second and cross indexed in limitless ways, your creative quests can gather about all of the world's apparently relevant information to feed your thinking.

You will undoubtedly learn about potentially mutually helpful people in your Internet explorations. These Other Minds become a collective intelligence community, especially if they are organized through an association, social media sites, Web chats, meetings, conversation groups in LinkedIn, or as an Internet group called a wiki.

The flow of information from the computer clouds, other minds, and your mind converges at a thought vortex that almost has a state of being and mind of its own. This vortex of information will underlie the solution of many very complex problems.

The Age of Google

Google is ever closer to the point of storing all of the world's open literature, papers, documents and website contents. This was bound to happen once the software algorithms and computing hardware became available and commercially feasible.

The Google search algorithm, called PageRank, was developed by Google founders Larry Page and Sergey Brin when they were at Stanford University. They started work in 1995 and had a prototype named Google in 1998. A patent was issued and assigned to Stanford. Google received exclusive rights to the patent in exchange for 1.8 million shares that the university sold in 2005 for $336 million.

PageRank ranks a page by the number of other pages linked to it and by the ranking of those pages. So a page with three links to it is ranked higher than a page with one link, assuming the other pages connected by those links have equal rankings. If the other pages' rankings increase, so does the ranking of the pages they are linked to. There are other undisclosed adjustments to the PageRank algorithm to prevent manipulation and other problems better known to Google.

The major plus for the Google PageRank algorithm is that it works well for rapid searches and with a relative minimum of user computer capacity required. Further, it does not require humans to adjust the page rankings directly within the Google database itself. The rankings are adjusted automatically whenever a person links one page to another.

The rankings for all pages in the Google memory change continuously much like the human brain continuously changes the degree of importance connections between subjects based on the continuing inflow of new information. For example, your brain has some relationship established linking "dogs" and "fleas." The negative for Google's PageRank is that

PageRank does not know what dogs or fleas are or the relation-ship between the two. It will display ranked pages talking about dogs and fleas if you tell it that both are important by typing in the search space "dogs fleas".

All of the Google indexing and search tools would be of little practical use except that computer technology now allows storing and accessing just about all the information in the world. The big development was realizing that huge super-computers are not the optimum way to handle the job. Small computers, not too different from a home PC, are networked together in any number and in many locations to collectively do whatever very high volume computing is required. These are called servers, and as of 2010, Google had 450,000 servers, which collectively consume over 50 megawatts, the equivalent of about 40,000 American homes.

As for throughput and storage capacity, Google processes about 20 petabytes per day where one byte is about equal to the information in one keystroke (letter, number, space, etc.) and peta means one quadrillion (10^{15}) bytes. It is estimated that this computer cloud collectively stores about 150 petabytes of data. Everything is programmed and arranged so that the elapsed time from your query to the result is within two-tenths of a second.

What can you do with that storage capacity? They have indexed and can instantly access over a trillion URLs (web-sites) and this pile of data grows at the rate of several billion pages per day. There is a copy of the entire Internet within the Google computer complex so that it can find and analyze information faster. Now Google is copying all the world's books to which it has been given access—which is probably the large majority—and it is three dimensionally mapping all the earth's surface to better than one meter resolution. This fits within the scope and spirit of the company's mission "to organize the world's information and make it universally accessible and useful."

It appears that the Google cloud is basically in place and from here must grow to accommodate the ever-increasing pile of information building up all over the world. According to an interview Peter Norvig, director of research at Google, gave to *Technology Review* magazine, Google search development is heading towards all kinds of content and interface modes. This can be seen now with the user-controllable view of almost any part of earth through Google Earth and with viewing of video documents stored in Google, such as the familiar YouTube.

These visual media have in turn multiplied the information storage requirements of the Google cloud by what Norvig estimates to be a thousand times. Not only is "a picture worth a thousand words," but compared to alphanumeric data, the requirements for transmission capacity through wires, fiber optics and satellites is very much greater.

The Search for Patterns in Big Data

Some mega data projects are focused on searching for patterns and empirical models by analyzing massive amounts of data. One is the Cluster Exploratory (CluE), a National Science Foundation program using Google and IBM computing platforms and tested in conjunction with six pilot universities. The cluster consists of 1,600 processors, several terabytes of memory and hundreds of terabytes (hundreds of trillions of bytes) of storage along with the software including Google File System, IBM's Tivoli, and Google's MapReduce. Early projects are in the biological sciences and include analyses of the brain and nervous system.

IBM has announced a new software package to find trends in large data sets. It rapidly analyzes huge volumes of data to find trends. Finance, health care and space research are among the applications IBM sees as good candidates for its new software. This software is the key offering of a new IBM division called Analytics.

As of late 2009, it has 4,000 "consultants" with a target of settling at 10,000-15,000 consultants. IBM Analytics has its own massive supercomputers. Its first target market is "Smart Cities" where IBM will help clients find ways to save energy such by better management of automobile traffic. In a 2009 interview, Sam Palmisano, then CEO of IBM, said that Analytics could help the world avoid a financial crash like the 2008 catastrophe by instantaneous analysis of data streams coming in from various financial institutions, primarily in New York.

If the cluster exploratory approach works, where are we headed from here? Stephen Baker in *Business Week* writes:

> What will research clouds look like? Tony Hey, vice-president for external research at Microsoft, says they'll function as huge laboratories with a new generation of librarians—some of them human—"curating" troves of data, opening them to researchers with the right credentials. Authorized users, he says, will build new tools, haul in data, and share it with far-flung colleagues. In these new labs, he predicts, "you may win the Nobel prize by analyzing data assembled by someone else." Mark Dean, head of IBM's research operation in Almaden, California, says that the mixture of business and science will lead, in a few short years, to networks of clouds that will tax our imagination. "Compared to this," he says, "the Web is tiny. We will be laughing at how small the Web is." And

yet, if this "tiny Web was big enough to spawn Google and its empire, there's no telling what opportunities could open up in the giant clouds."

What of the Scientific Method?

Some people say that the era of big data and development of patterns by correlation of data sets will make the scientific method obsolete. Chris Anderson of *Wired Magazine* wrote:

"All models are wrong, but some are useful." So proclaimed statistician George Box 30 years ago, and he was right. But what choice do we have? Only models, from cosmological equations to theories of human behavior, seemed to be able to consistently, if imperfectly, explain the world around us. Until now. Today companies like Google, which have grown up in an era of massively abundant data, don't have to settle for wrong models. Indeed, they don't have to settle for models at all...The big target here is...science. The scientific method is built around testable hypotheses. These models, for the most part, are systems visualized in the minds of scientists. The models are then tested, and experiments confirm or falsify theoretical models of how the world works. Science has worked this way for hundreds of years.

It might appear that the scientific method has become a charming anachronism like the Dodo or the slide rule, but it is not a question of "either" or "or." It is a combination of "both." Pattern analysis in massive amounts of data should not be thought of to produce a final model for scientific purposes.

It should be used to produce a working hypothesis based upon which an analytical model is developed. It is like producing sketches before composing a painting.

A key point we should never lose sight of is the quality of the data. Many believe that the highest quality, most reliable data comes from direct observation. This is in contrast to deriving information from searching for patterns in big data.

Naturalists, going back to Darwin or earlier, have long stuck to their discipline of making extensive field notes every day. My cousin, Ben Kilham, who is a world authority on bears, tells me that through his daily field notes observing bears in the wild, he is constantly accumulating information not found in any computer repositories.

Now drug researchers, who are confronted with collections of millions of data points, are rediscovering the meaningfulness and productivity of observing individual molecular reactions. They see unexpected reactions, which may be the key to new drugs. Individual home workshop inventors may still profit from hundreds or even thousands of apparently meaningless experimental variations in order to fully understand the materials and designs of their latest inventions.

Implications for Inventors

Despite the easy use of the computer clouds and the proliferation of communications devices to connect with them, creation and invention have not been automated. Successful creation of major inventions requires as much as a lifetime's pursuit of relevant knowledge, full use of the Web, experimenting with possible solutions in a hands-on laboratory, informally tossing ideas around with real people in one place, and a lot of solitary thinking. That is why there will always be room for the Edisons and the Leonardos.

My friend Dr. Joseph R. Stetter, a chemist with more than 25 patents and many awards for creativity, offers this interesting perspective as we enter the Age of Google:

> There was a time when I spent time each month in the reading room in the chemistry library browsing my favorite journals. There was a time when I visited the library on the spur of the moment whenever I had an idea to elaborate, or a problem to solve, or any issue for which I needed added information. There was a time when I took and later taught courses in the organization and searching of the literature of science and engineering.
>
> In brief, invention is "building a better mousetrap" and innovation is "doing something different with the invention for social impact." Of course, my career modus operandi in science and engineering would have evolved in any case and in many areas of the scientific endeavor since progress in scientific instrumentation as well as knowledge has evolved rapidly and significantly.
>
> But, in the aspect of scientific invention and innovation, no greater impact has been observed than that of the Internet and allied technology (search engines, content providers, etc.). No longer do I need to go to the library to find information. And while this sounds simple, it is an immense transformation.
>
> A good research group or research university was judged in part by how good the library was. Now stature in research capability is judged by how good your search engine might be. I still love libraries but it is a nostalgic love and not a love from need. I love the atmosphere of a library and the feel of a book. However, I now have at

my fingertips, still for a subscription fee of course, multiple libraries that grow in quantity of information with unbelievable rapidity.

My challenge now is to understand the search engines, so I am not mislead to what the provider wants me to see and to the best information to understand my issue. Operationally, I now first go to my computer when I need information rather than the library. The course I last taught in chemistry literature was not so much an emphasis on library literature and libraries but about distinguishing the quality of information from anecdotal web page postings to the highest quality refereed journal articles and the ongoing scientific discussion that improves our theories and knowledge base.

Finally, I absolutely enjoy having information from technical to social immediately available to solve a problem and find it enabling of invention and innovation on many fronts. The ideas that I have come to me at odd times, and often can only gel if they can be incubated quickly. This is now realized and I look forward to the day when every piece of literature in every language is available immediately to every person on earth on their cell phone! What a transformation of thought, operating style, and creative stimulus we have seen in less than a single lifetime!

11

COLLECTIVE INTELLIGENCE: MANAGING LARGE-SCALE RESEARCH AND INNOVATION

We as a species have entered a new phase of evolution with the appearance of the World Wide Web. You can find out almost anything you want to know at the click of a button, and this happened suddenly, nobody predicted it. This is a collectivization of human information. Once you start to act with other people, you can do things you couldn't do as an individual. You become a connected intelligence and just like joining computers together, that increases your effectiveness and power... For scientists, it means the world is now one giant research group.

John Barrow, *The Infinite Book*

A fundamental mechanism of human intelligence is combining known and unknown data. This is a process of increasing your knowledge and often increasing the knowledge of other like-minded people as well. This goes back to the cavemen who sat around the campfire and shared ideas about which spear head worked best on the hunt. This seems to be the case with more elementary creatures as well. Adjacent members of a species are generally aware of each other's presence and will share concerns if they can. This is known among biologists as "quorum sensing."

With the advent of Internet forums, wikis, affinity groups, blogospheres and so on, we humans have embraced much broader forms of problem solving and innovation. This learning and group communication mode is becoming known as "collective intelligence" and "crowd-sourced innovation." In its most general form, it is determining the consensus of many minds to find a response to a complex challenge. For example, collective intelligence could be used to find solutions for many problems engendered by climate change.

Examples of Collective Intelligence

The MIT Center for Collective Intelligence has been shepherding the development of an online forum called the Climate Collaboratorium. It will be a constantly evolving computer model of the Earth's atmosphere and human systems with inputs from online scientific chat rooms. All the variables and factors that can be imagined relating to climate, the environment, interactions with human beings, and ecology are included in the evolving model.

Professor Thomas W. Malone, the center's founding director and MIT Professor of Management, compares the Collaboratorium to the Manhattan Project, which developed the atomic bomb during World War II. "The difference between the Climate Collaboratorium and the Manhattan Project is that this is a problem everyone in the world needs to solve, but because of new technologies like the Internet, it's possible to enlist far more people than during World War II."

Malone's climate project has been formalized to the "Climate CoLab." As of late 2014, it has over 15,000 registered members from more than 150 countries. Their organizations include NASA, the World Bank, the Union of Concerned Scientists, leading universities, businesses, government agencies, and student organizations.

There are of course many examples in history of collective intelligence that has gone wrong. Contrary to the then prevailing wisdom we now know that the world is not flat and it is not at the center of the solar system. Wars have been started by propaganda whose content is skewed or falsified to inflame emotion.

There are however plenty of examples that could be cited to show the amazing accuracy possible with collective intelligence. Malone and his associates at the MIT Center for Collective Intelligence report:

- *Kasparov v. the World* was a chess match held in 1999, when world champion Gary Kasparov played against "the world," with the world's moves determined by majority vote over the Internet of anyone who wanted to participate. Kasparov eventually won, but he said it was the hardest game he ever played (at least until he met IBM's "Big Blue" super computer and lost).
- *NASA Clickworkers.* In 2001-02, NASA let anyone look at photos of the surface of Mars on the Internet and identify features they thought were craters. Crater locations were designated by sets of coordinates in two-dimensional space. When the coordinates contributed by amateurs were averaged, they were found to be just as accurate as the classifications made by experts.

The most common example of successful collective intelligence is the Google search engine. Its search for the ranked results of any inquiry is based on the collective intelligence drawn from all Internet searches to the present. The collective intelligence is the number and locations of linkages between websites that were developed in prior searches. Google remembered these. It does not know the expertise of the people who made those prior searches. Some people argue that this is a weakness of the Google algorithm and others argue it is a strength of the algorithm.

Malone and colleagues divide organizational goals of collective intelligence groups into:

1 - *Create.* The participants create something such as an entry in Wikipedia, more software code for Linux, or a poem in honor of a member.

2 - *Decide.* The participants evaluate and select alternatives such as estimating climate change effects next year or deciding whether to delete a Wikipedia article.

Because we are concerned with innovation and creativity, I will focus on the "create" aspect of collective intelligence. First, we should agree that the inventor or creator when researching a particular subject sooner or later must engage collective intelligence. In the old days, a corporate researcher might consult the books, journals and papers in the company library and exchange ideas with his or her colleagues in the company cafeteria.

This procedure has been replaced or at least augmented by searches for relevant information in the Internet, searches in the company's own computerized database and its wiki if it has one, and idea sharing with colleagues by email, blogs, the company wiki, etc. The volume of source information is orders of magnitude more than the old days, the speed and flexibility of access is unimaginably higher, and the number and distance of people who can be involved in the creative process is no longer limited.

The Boeing 787 Dreamliner Project

An appropriate case in point is Boeing. For commercial jet aircraft, this company is noted for many innovations such as the first enduring airliner, the 707; the largest aircraft in the world, the 747; and the most purchased model, the 737. In 1995, Boeing started shipping the 777, a long-range, wide body jet and the largest twinjet in the world. For the design of this plane, Boeing worked with its long-time software partner Dassault Systèmes to model each of the plane's 10,000 parts in 3D on the computer instead of building physical prototypes.

The 787, which followed the 777, moved collective intelligence design into mid-stream. "We had over a thousand of our partners' engineering personnel here to jointly define the airplane," said Mike Bair, head of the 787 program for Boeing. "That way we get the best ideas from everybody, as opposed to just ours."

The 787 Dreamliner, in commercial service since 2011, in many ways has been a visionary great leap forward. It was organized to use the 777 virtual design approach and to exploit the advantages of using composite materials for approximately 50% of the primary structure instead of aluminum. The composites are carbon fiber reinforced plastics. The features that are to be realized from the use of the composite materials include:

- 20% less fuel consumption due to lower weight.
- Better cabin air quality because the higher strength of the composites allow greater pressurization (the equivalent of 6,000 feet altitude instead of the normal 8,000-foot

equivalent pressurization) and higher humidity because the composites will not corrode from humidity condensation like conventional aluminum.

- Fewer parts, such as 50,000 fewer fasteners, due to larger one-piece fabrications replacing smaller plates fastened together.
- Less assembly time.

In order to accelerate the design at lowest cost for this innovative plane, Boeing decided to shift much of the design to its supplier partners. Previously, Boeing designed 70% of the aircraft. Now its 43 top-tier suppliers and many more sub-tier contractors from 24 countries would work at 135 partner sites. Key to project progress was getting these design-partner suppliers to give up their individually favored computer aided design (CAD) systems for the common language and format of Boeing's Catia V5 system.

The Global Collaborative Environment (GCE) networks all members of the 787 design team. In addition to the Catia CAD design program used by all design members, the Dassault suite of program management tools include Delmia for digital manufacturing and Envoia to inventory all information on the 787.

Once this was done, the supplier could design many of the parts or assemblies from more general specifications supplied by Boeing. The communication between the supplier and Boeing was enhanced and standardized using the same engineering design and data communication programs. As a result, specification to a supplier dropped from 2,500 pages to 20 pages. Mike Bair explained, "We've realized that it's more effective when the people who are building the parts also do the engineering. They know better than us how their factories run, and to think that we can design a part that not only serves our needs but is also the most efficient for them to produce would be pure guesswork on our part."

While using vendors in a pool of collective intelligence to help design your product through customization of their designs for you is not a new idea—I have used this approach as a central focus in several of my technology companies—what I can attest to from hard experience is that doing it is much harder than describing it. What makes this approach possible to a much greater extent today is massive database and networking computer technology and the new generation of engineers' positive attitudes towards the use of collective intelligence.

It could be said that the Boeing team design approach is more collaborative than collective. The collective approach might assume more open-ended outside contributors and more freedom of ideas on their part than simply finishing designs whose general guidelines came from the project's management.

There are at least three reasons why Boeing's approach is less than fully collective. One is that once a collaborative effort is launched, the company is not looking for fresh, new designs. The basic designs and parameters have already been determined for each part or subassembly.

In addition, the company keeps new ideas, concepts and designs proprietary. All intellectual property belongs to Boeing or its collaborators. Enforcing control of intellectual property is much more difficult when contributors can come and go at will. Eventually controversy can develop about who owns which ideas.

Finally, collaborative intelligence such as can be attained through a wiki makes proprietary and secret information impossible to control. I know of a national government laboratory where the scientists are very frustrated about the lack of a computer-based bulletin board or chat group through which to post data and exchange ideas. The challenge has not been solved about how to prevent the escape of national secrets.

Solutions from the Internet Challenge

Another approach to innovation and design using the flexibility and outreach of the Web is to broadcast a request for an invention or design to anyone who cares to rise to the challenge. This is a good example of crowd-sourced innovation. There is a reward to the winner of the challenge—usually in the 10s of $1,000s. A leader in offering this type of service is InnoCentive based in Waltham, Mass. It is a global Web community enabling scientists, engineers and others to collaborate to deliver breakthrough solutions for innovative R&D-driven organizations. Since 2001 their clients have included Procter & Gamble, Eli Lilly, The Rockefeller Foundation, government agencies, and non-profits.

As of early 2009, 814 challenges had been posted (a "challenge" is a request for an innovative solution), for which 12,529 solutions have been proposed, and 391 were accepted and received cash awards. Among the challenge solutions InnoCentive feels are of broad significance are:

- *Oil Spill Recovery.* In 2007, the Oil Spill Recovery Institute posted three challenges dealing with oil spill recovery. One of these was solved later that year by a researcher who proposed a solution based on his expertise in the concrete industry. Insights can be found unpredictably in unrelated applications.
- *Towards eradication of tuberculosis.* Also in 2007 the TB Alliance, a not-for-profit development partnership dedicated to accelerating the discovery and development of

drugs to treat TB posted a challenge on the InnoCentive web site to simplify the manufacturing process of a current drug. Solutions were provided by solvers, one of whose mother was a victim of the disease.

- *Clean water in Africa.* A water filtration system has been developed that uses carbonized coconut shells to filter out large particles and heavy metals. An ultraviolet LED powered by a solar panel then sterilizes the water.

Considerations When Using Collective Intelligence

Collective intelligence is better for idea generation than idea evaluation according to Eric Bonabeau writing in the *MIT Sloan Management Review*. He points out that management decisions today must be made more rapidly in the face of much more data and opportunities than was the case of our ancestors. Our limitations as individuals may not be sufficient for today's decision-making, and perhaps we should rely more on others to find solutions. Outreach is needed to obtain diversity of stimuli, assumptions and solutions.

For many problems, a solution can be found outside of the company, so tools need to be developed for tapping into helpful outside sources. Understanding must be used to balance between diversity vs. expertise and decentralized vs. distributed decision-making. Care must be taken to tap the crowd's wisdom and not its madness.

Now the question arises: where would the visionary thinker, the Thomas Edison, fit into this entire comforting new world of collective intelligence? Edison did the best he could without the benefit of Web-based collaborative intelligence, Google or wikis. He achieved much the same thing by surrounding himself with an army of engineers, technicians, and lesser-known inventors and by maintaining his own large technical library. He also gained venturesome technical insights from his social associations with such industrial entrepreneurs as Henry Ford and Harvey Firestone.

Many people would consider all of this prohibitively expensive and time-consuming to do today. They would reach for their laptops and see if solutions could be found there.

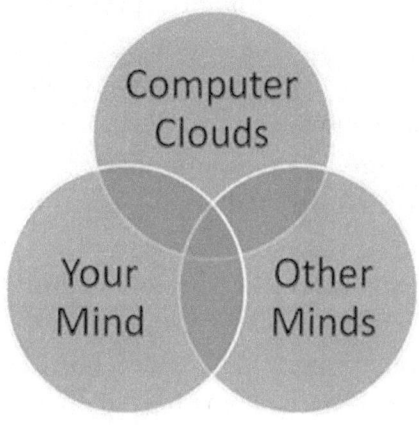

Where will you position your mind?

Leonardo da Vinci, were he alive today, would join Internet chat groups and log into the computer clouds and addition to keeping a close circle of friends at his workshop. He would eagerly pick from the nearly infinite number of creative challenges that could be presented to him today. Leonardo would agree that with our Web incorporating billions of websites and quadrillions of pieces of information, with access to all by a laptop computer, that this data and computing cloud is closer to his infinite universe than simply being able to see the outer limits of the actual universe.

Leonardo's practical nature would appreciate that the use of modern communications technology and the Internet is more efficient, and he would not waste time on inventions that already exist.

Google and the other computer clouds are a new paradigm of creative thinking. The complex and critical problems confronting civilization today are demanding solutions at a much less leisurely pace than in the past. Darwin could arrive at his theory of evolution over a lifetime with no contrary consequences of his slow, methodical progress. Today, however, problems including global warming, food and water, and the tightly coupled national economies require genius level insights in a short time. For a particular problem, if the right solution is not agreed upon and enacted in time, the world system may have crossed the tipping point beyond which the system cannot recover.

As the predominance of the lone inventor gives way to highly computerized research labs and teams, concern mounts about data management. There has been recognition that stewardship of data in the era of big data is a major priority. In January, 2009 the White House National Science & Technology Council issued a study, "Harnessing the Power of Digital Data for Science & Society" which noted that "preservation of digital data is both a government and private sector responsibility and benefits society as a whole."

CONCLUSION

THE FUTURE OF MEGAMINDS

The problems of the world cannot possibly be solved by skeptics or cynics whose horizons are limited by the obvious realities. We need men who can dream of things that never were.
John F. Kennedy

Planet Earth is spinning towards a new intellectual ecology. Due to massive low-cost computer clouds and nearly limitless communications networks connecting everyone, culture will irreversibly change and will change the way everyone lives. The brain that made man special over all the other creatures has created a networked brain about whose magnitude and consequences we can only speculate.

This growing Internet brain will offer any kind of instant data and apparent solutions to problems. For humans to maintain their independence, however, personal programs of achievement and education must emphasize that truth is the important goal of the searches and not feel-good satisfaction. Creativity and invention will follow.

A pioneering proponent of the worldview was Pierre Teilhard de Chardin (1881-1955), a visionary French philosopher and Jesuit priest. He was trained as a paleontologist and geologist and was involved in the discovery of Peking Man. Teilhard de Chardin saw a sheath of consciousness surrounding The Earth, which he called the Noosphere (from the Greek, literally "the Mind Sphere," or the sphere of human thought). This was a "planetary thinking network" of interlinked consciousness and information. Awareness of it would increase with ever more complex social networks. Key outcomes would be a global network of self-awareness, rapid feedback, and communication.

While Teilhard de Chardin did not live to know the Internet, in view of his prescient ideas, he must have sensed that something like it was coming. He felt that with the emergence of the Noosphere, the age of nations will pass and The Earth will replace The Nation. I refer to today's version of the Noosphere, which is an infinity of computer clouds, as the Knowosphere.

As time passed, the sense of euphoria about the Planetary Thinking Network has turned to a colder sense of impending reality. Susan Greenfield, a neuroscientist at Oxford University, wrote in *The Guardian* on February 10, 2005:

> You're just a consumer, living at the moment, having an experience, pressing buttons but not having a life narrative anymore. You're not defined by your family, or by what you know, or by specific events in the real world, because most of your time is spent in cyberspace. So what are you? Could it be that we just become nodes on a much larger collective thought machine?

It is easy to settle into being a cyber traveler, but the journey can get you ever farther from the focus on truth. In order to make useful sense out of the masses of the data in the computer clouds, you should constantly check the accuracy the answers you receive and the models you build.

There is a need for a new kind of thinking in the face of the recently available mountains of data—data instantly accessed and conveniently packaged like a supermarket consumer product. In order to break loose from a steady diet of packaged information, you must fire up your imagination and embrace new ideas. From that start, creativity and innovation leads to new frontiers, and education is the basic force for insuring productive change from generation to generation.

The Individual, the Internet, and the Truth

The creative person seeks something more than being part of a computer cloud-based social affinity group. He or she increasingly will have to balance plugging into the computer clouds and collective wikis on one hand and continuing the search for the unexpected that may not be in the clouds on the other hand. There is the balance between the insightful mind and the collective intelligence. There are the different worlds of the instant Other Reality on the Internet versus the very long private intellectual quest. There is the abstract thinking mind versus the ever-improving machine learning.

Therefore, the creative person seeking new truths and models must constantly try to sort out the relevance of the computer clouds information versus what he or she thinks independently. People immersed in information analysis must increasingly be on guard that rumor and invective may be outpacing the seeking of truth.

The industrial revolution has given way to the knowledge revolution. The evolving mind of man may be overwhelmed by the computer clouds. The combination of peoples' minds and clouds' minds can be thought of as a super mind. A growing problem however will be that our thinking may not be so focused towards complete analysis and the best analytical methods. Holistic solutions to problems perceived in mega-data sets can lead us to new insights, but they can also lull us into overlooking reductionist analyses and model building typical of the classical scientific method.

Noted MIT computer scientist Joseph Weizenbaum (1923-2008) was not impressed by the Internet and was troubled by the fact that it was easy for people to mistake pattern matching for true understanding. For the individual human trying to make the most of the Internet, maintaining perspective about the apparent truth emerging from the data analysis will be very difficult. One way to do this is by seeking perspective through original research.

All over the world people are sensing that knowledge and information have become a common resource and tool, and that has led to whole new Web services, "social media" such as Facebook and Twitter, and Web-based mega companies such as Google and Amazon. Collective thinking, however, can lead to mantras based on questionable science. People can all start thinking that honeybees are dying because of insecticides; and when reinforced by Internet chatter and talk shows, the group diagnosis becomes The Truth.

Therefore, anyone finding newly discovered "scientific" revelations and discoveries on the Internet should be skeptical. Scientific detachment would indicate wait for some time to pass, seek other opinions, and search for probing questions to test the new and often popular theory. Use the Internet and other resources to find opposite points of view or broader perspectives about the issue.

Imagination and creativity, integrity, a sense of wonder and truth, persistence of inquiry—these keep us above the turmoil and herd mentality of the Internet. They allow us to profit from the inexhaustible information resource of the Internet if we keep our sense of perspective and good judgment.

A New Awareness and Education

Everyone should know that we are in a paradigm shift with respect to knowledge acquisition and understand that there are serious consequences for sloppy data interpretation. Otherwise, we could be in for an era of increasingly misunderstood big science and misapplied analytical technology at the enterprise level.

I woke up to this emerging reality recently when judging an elementary school science fair. This school was serving the poorest part of town with many of the students being the children of undocumented Mexican immigrants. Nevertheless, they were bright eyed and anxious to show their exhibits of such things as the way plants grow with different nutrients and demonstrations of gravity and motion.

As I talked to the kids, it dawned on me that they were getting a lot of their project ideas and help from the Internet. Some kids were getting their project designs from Web-based services catering to science fair students. I estimated that 20% of those in grade two and 80% in grade six were doing this. There is a warning sign here that the unconditional use of the Internet packages may have proliferated due to an emphasis on right answers and attractive demonstrations with a disregard of the science involved. On the positive side, maybe the Internet set the kids' minds in motion, and perhaps they will embark on fruitful scientific careers.

About that time, I saw an article by Monica Hesse in the Washington Post where she asked, "For the Google Generation, what happens to the concepts of truth and knowledge in a user-generated world of information saturation?" She reports on a college freshman who never checked out a book from the university library, but who was overwhelmed by all the Internet information. The student remarked, "The idea of having an original thought terrified me." Once she realized how much information was out there, the idea of productively using it seemed impossible. Part of ignorance is not knowing how to filter out irrelevant information.

We must inculcate the value and methods of good research in our student population. As they enter the enterprise world, be it public or private, profit or non-profit, they must have and maintain a critical attitude towards information, knowledge, truth, creative ideas, invention, flights of fancy and imagination. They must realize that the pursuit of new scientific insights must include focusing on finding truth among all the information and not just processing the information itself. Without the sense of importance of empirical truth, the relevance of reality is lost and progress is stalled.

The Role of Education

Again, my father has something important to say. He was willing to give education a chance, but became discouraged. In 1933 he wrote:

Now let us look at schools and see what the future of the country may be. Are people made like machines, in quantity and to a set standard? Or is education a means of developing the individual? In the main, I think we shall find schools to be factories of the more expensive variety. Of course, there is at present a very strong trend in teaching towards broader and more individual education, but in general, what have we? Classes of numbers of pupils, I believe, all of whom are graded in a series of numbers that do not relate to any of them. Classes in which the book is the thing, not the pupil. Classes where memory ranks high, imagination low. Classes where every subject is divided into parts instead of classes where the parts are combined to show the whole. Self-sufficiency demands a broad general education put to use according to the need. Schools give a pigeonholed education almost totally removed from use. They tell us it is "mind training" but the memory is only part of the mind.

I could not agree more. It was the same in my education 20 years later.

My father, Peter Kilham, surrounded by bird feeders
he invented. Foster, Rhode Island, in the 1980s.

Formal education is very important for entrepreneurs, but as has always been the case, many of those with creative minds tend to be impatient and often drop out of school before they complete their formal education. They become frustrated with the formality and rigid structure they perceive to be endemic to classroom education. There is a tension between the accumulation of facts and nurturing creativity. There does not seem to be any room left for the mind to wander, catch a glimpse of a new vision, or pursue it wherever it may lead.

In the twenty-first century, the quantity of information is growing so rapidly that it is impossible to select what should be taught to students as the basis of The Knowledge that they will need in the future. Besides teaching the basics, schools will have to focus on providing students with the skills they need to solve problems and questions that cannot even be asked yet, let alone answered.

Correspondence schools and community colleges traditionally have been a source of second start education for those who grew to realize that they did not learn enough to achieve some aspiration. Today's correspondence courses and some community education courses are offered with the convenient and appealing media of online transmission. Some universities including MIT are offering their courses free online to anyone who logs in to them. These are called MOOCs (Massive Open Online Course). This is a definite second chance for some self-directed individuals.

Where possible in such cases however, it would be very helpful to have a coach or tutor critique and guide the otherwise self-guided student as they try to sort out this stream of knowledge. With highly interactive computer clouds offering multimedia education in a reality-based dialogue method, the students could find themselves in a virtual classroom with personal attention.

When students become employees in industry or government, they often will find more interest in new ideas than seemed to be the case in schools, especially if they are employed in technical areas such as engineering. However, deficiency of essential formal education often shows up as lack of essential technical knowledge or communication skills.

I have encountered many entrepreneurial technicians and engineers who hit a brick wall because they did not know the physics or chemistry involved in their inventions. It is very difficult to catch up in deep technical areas later in life. They should have studied more science and math in schools and universities. The areas of significant technical invention today usually are much more complex than in Edison's day, so prospects are much dimmer for the essentially self-taught entrepreneurs. Equally, a stumbling block is the lack of communication abilities on the part of these entrepreneurial hopefuls. They cannot seem to explain in understandable language what they are thinking or proposing. They cannot read published information that is required to support their project. They cannot write down their findings and notes for their associates and followers.

Our schools apparently have the reference resources students need in terms of both technical education and communication skills, but this knowledge often does not seem to be getting through to the students. Three things need to be done:

1 – Get children interested in creative accomplishment at an early age and keep them focused on this throughout their lifetimes. This requires teachers who love what they are doing. Teachers who are on fire. Teachers who love science and really want their students to absorb it.

2 – Make sure that the fundamental knowledge needed has been presented and learned. If teachers do not know their course material, replace them with ones who do.

3 – See that the students who are interested in innovation, invention and entrepreneurship do not drop out of school prematurely, foregoing the additional technical education and communications skills that they will need.

We will need many graduates who are hooked by the challenge of the unknown. They must be captivated by the wonder of unknown and the goal of making a unique contribution to its understanding.

This motivational process starts at the top—with the President of the United States—and carries through political and business leaders, parents, clergy, educators and many others. When Russia launched the first orbiting satellite, there was frenzy in the United States not to fall behind in the technological race. We put our man on the moon first, and this goal has faded out. Now the world is faced with larger and sometimes irreversible problems of environment, climate, food, water and energy, and a new sense of mission must be developed.

Towards the New Frontier

Times of insight and creativity come and go with the ebb and flow of unexploited knowledge and with society's sense of urgency for new solutions. The industrial revolution and World War II were eras that saw surges of insights, creativity and invention. Now the world is benefitting from a combination of bright new minds coming up through the educational systems, well-equipped laboratories and shops, and the new information sources of the computer clouds; but there is an apocalyptic sense of the world running out of time. In addition, many people feel a sense of "Why bother?" because it appears that the world has run out of possibilities.

People must see that the whole universe is available to them and that creativity has never been more important than now. Children should realize that there is an infinite future for them. Society's failure is failure to give them hope and encouragement.

Now is the time for the men and women who dream of things that never were. Their dreams are the starting points in great creations. The positive emotions of the challenge will cause the complexity and depth of the world's problems to fade away. The one catch is that their dreams will have to answer to unmet realities.

It is time to turn America and the whole world into a nation of creators and inventors again and for the whole world to work together to deal with the many challenges and opportunities that are upon us. From garage inventors to multinational corporations we must make a fresh effort at creativity and innovation including using the vast new resources of the Internet and the computer clouds. America and the whole world need to become more creative in all endeavors.

The Great Summing Up

In the waning months of his life, I was with my father in his combination shop and office. He was looking through his company's yearly financial reports, noting profitability increases every year, and looking at me for a response. I did not know what to say or how to respond. They were just sheets of paper with numbers on them being shown by a great inventor. He did not refer to his inventions, models or artwork. We fell silent and then slowly left for the kitchen.

Over the years, this scene has haunted me. The same questions have kept plaguing me: What was his purpose? Did he want me to take over the business? Was he disappointed in the way his life turned out—thinking of himself as becoming merely a shopkeeper? Did he think that the financial statements were all that I, a business school graduate and corporate manager, could understand? Was I incapable of understanding him?

As I write this book, the answer comes. The financial figures were the distillation of truth. It had taken a lifetime, but the world at last had come to accept his creations and was willing to pay for them. Not only had he had created useful and beautiful things of lasting value, but they were highly valued in the market place. He had every reason to be proud.

Unknown to me, he knew this conversation would be about our last. He was saying, "Be guided by the truth, and the rewards will come."

GLOSSARY

Algorithm. A set of instructions to carry out carry out a procedure or process. It is usually in the form of a computer program derived from a set of equations or a flow diagram.

Artificial Intelligence (AI). The ability of a computer or other machine to perform those activities that are normally thought to require intelligence. Artificial intelligence devices range from bug-like robots that can find their way around a terrain to mega-computers that someday may be more intelligent than humans. The term artificial intelligence is used in a variety of ways ranging from an engineering and scientific discipline to a specific computer, program or device.

Associative Inference. Calculating or inferring the strength of the connection between two words or data points in a knowledge network such as a computer database or a network of neurons. While all points are usually connected to each other, the strength of any given relationship is determined by the sum of the strengths of the direct and indirect connections between the two data points. Example: "dog" and "ice" would not be likely to have a strong connection, but "dog" and "fur" would have a very strong connection based on direct and indirect (multi-linkage) connections.

Autism. A brain development disorder characterized by limited social interaction and communication. Autistic people seem to think more in terms of pictures and numbers than in verbal communication. Many autistic people have nearly perfect memory in both detail and the amount of information that can be remembered. Autism may provide some clues about the ways the brain thinks and imagines. The causes and treatments of the condition are still not well understood.

Bipolar Disorder. A psychiatric disorder characterized by periods of mania alternating with periods of depression. There is a theory that all very creative people are to a certain extent bipolar.

Bytes. One byte represents the memory capacity in a computer to store a letter, digit, character or space. A **kilobyte** (1,000 bytes) is about the storage requirement for a page of text. A **megabyte** (1,000 kilobytes) is about the storage capacity required for a medium resolution photograph. A **gigabyte** (1,000 megabytes) is about the storage capacity of a CD disk. 100 gigabytes or more is the storage capacity in a PC or laptop. A **terabyte** (1,000 gigabytes) is the size range of very large storage in corporate and government computer systems. A **petabytes** (1,000 terabytes) is a suitable dimensional unit for a computer cloud. The Google cloud represents well over 100 petabytes.

Cartesian. Often referring to Descartes, and his mathematical methods and philosophies. More generally, usually means emphasis on logical analysis and its mechanistic interpretation of physical nature.

Clouds, Computer Clouds. An evolving term referring to computer services accessed in the Internet, especially where there are extremely large capacities of programs and data storage. The user generally does not need to know about the technology of the computers, software or data handling.

Cognitive Information processing. Computer theory and devices designed to replicate cognitive thinking such as done by humans. A device has sensory inputs and short- and long-term memories. Much of the development in this area is done with associative inference and neuronal chips that are similar to arrays of synapses.

Collaborative Intelligence. People working together to solve a problem, accomplish a design, or other action generally too complex for a lone individual. The working environment may be in a room, by way of the Internet or through a wiki. For collaborative intelligence, the people involved know each other as associates, contractors or through other relationships.

Collective Intelligence. Similar to collaborative intelligence above except that many of the people involved do not know each other. People can join the group and leave so that the composition of the group may be changing all the time.

Connectionism. A construct in artificial intelligence modeling the mind as functional modules and nodes wired together. It is also used in cognitive and neural sciences to model the thinking process as the flow of information between interconnected units. The network nodes are identified in numerical terms.

Consciousness. Conscious is a very widely used word that, depending on the observer or situation, may have several meanings. Therefore, it will be forever difficult to define to everybody's satisfaction what consciousness is. Your consciousness could be said to be everything you yourself are aware of. This awareness is a synthesis of your past and present, taking into account all your senses and memories. Since you are aware of yourself, conscious includes self-awareness but it is not of itself intelligence.

Cryogenic. Relating to the production of very low temperatures.

Cybernetics. The science of control systems for living beings and machinery. Derived from the Greek word *kubernētē*, governor, cybernetic means the control of the organism or machine from a self-correcting point of view. It was a precursor to artificial intelligence but today it is not generally considered a central construct of artificial intelligence.

Database. A collection of data, usually stored in a computer, that intends to have its elements interrelated by one or more themes or relationships.

Data Sets. Data stored so that rows and columns represent particular variables. As the term is used in this book, data sets usually means meaningful data relations, hitherto often unknown, which can be found in massive databases.

Determinism. The philosophy that all actions are the consequence of prior actions and that these relationships between current and prior actions can be described analytically.

Dopamine. A neurotransmitter involved in regulating movement and emotion. Some think that an excess level of dopamine causes both attention deficit hyperactivity disorder (ADHD) and creativity. Geniuses may have high levels of dopamine.

fMRI. Electromagnetic imaging brain scan used to plot the locations of mental activity in the brain under various experimental conditions.

Hippocampus. A small structure found deep in the brain that is a key manager of the memory process. While the hippocampus is not unique to humans, it is considered helpful in advanced thinking processes such as sorting out what is novel. It is also a key brain component for managing spatial navigation.

Holism. The idea that the system cannot be determined by its parts alone. The term is often used to mean approaching a complex problem from the top down—looking for patterns in all of the data and then developing an analytical model. *Reductionism* is posed as the opposite to holism, and it maintains that a system can be determined by its component parts.

Intelligence. Like consciousness, intelligence is a word that has no single definition and its meaning is often more fully defined by the context in which it is used. In this book, I use intelligence to mean the mental powers to learn, reason and act and, most importantly, to grasp relationships and meanings and to think abstractly.

Knowosphere. All of the available electronically stored information in the world that is accessible by search engines or found through social media.

Manic-depression. Older term for what is today usually called bipolar disorder.

Moore's Law. A statement that technology tends to improve exponentially. Gordon Moore observed that improvements in miniaturization led to a doubling of the number of transistors on an integrated circuit chip every 18 months (variations of the statement range from 12 to 24 months).

Neocortex. An addition to the cortex called the neocortex. It is the outer layer of the upper hemisphere of the brain and consists of six layers. Its functions include sensory perception, the ability to learn with behavioral flexibility, motor commands, conscious thought and, in humans, language. In humans, the neocortex is 90% of the cerebral cortex up from a very minor percentage in simple mammals.

Neural Networks. Used to describe brains and artificial minds in terms of wiring or neurons connecting various data or information nodes. See *connectionism* above.

Neuron. A biological cell that is the fundamental storage and interconnection unit of the nervous system and brain. Neurons connect very rapidly with another with great precision through interconnection elements call synapses. The human brain has about 100 billion neurons and over a thousand synapses per neuron.

Prefrontal Cortex. Located in the front of the brain, the prefrontal cortex is especially prominent in humans and is associated with planning, decision-making, and higher level cognition.

Reductionism. A philosophical position that a complex system is the sum of its parts. It is the basis of classical analytical science. The opposite of reductionism could be said to be *holism* described above.

Robot. A machine that is controlled by a computer and sometimes by a person or other computer. It is often made to look like a human or animal. Robots generally fall within the category of artificially intelligent devices.

Schizophrenia. A psychotic disorder characterized by withdrawal from reality, illogical patterns of thinking, delusions, and hallucinations, and accompanied by other emotional, behavioral, or intellectual disturbances. Schizophrenia may be caused in part by dopamine imbalances in the brain. There has been conjecture of schizophrenia being a possible component of genius behavior.

Sentience. Feeling, sensation, absorbing external stimulation.

Sentient machine. Not precisely defined yet, but used in the media to mean a computer that has self-awareness and artificial general intelligence.

Serotonin. A neurotransmitter chemical in the brain involved with the regulation of mood states including depression, anxiety, perception and transmission of impulses between neurons. It has been speculated that serotonin has a role in creative thinking.

Singularity. A term made popular by Ray Kurzweil who has defined his use of it as: "The Singularity will represent the culmination of the merger of our biological thinking and existence with our technology, resulting in a world that is human but transcends our biological roots." (Ray Kurzweil, *The Singularity is Near*, Penguin, New York, 2005). In an interview about that book, Kurzweil said: "We'll get to a point where technical progress will be so fast that unenhanced human intelligence will be unable to follow it. That will mark the Singularity." Singularity is often described as the point in time when computer intelligence exceeds human intelligence. Kurzweil forecasted this point to be 2045.

Social Media. Media and tools for people to share information primarily by the Internet. Currently popular examples are Facebook and Twitter.

Synapse. Neurons connect very rapidly with another with great precision through interconnection elements call synapses. The human brain has about 100 billion neurons and over a thousand synapses per neuron. Synapses could be thought of as electrical relays or transistors, and their mechanism are chemical and electronic.

Synaptic Processor. Uses associative information processing with electronic synapses and massively parallel processing. Energy consumption is minimal. Used experimentally to do human-like computing. This is the alternate architecture to the conventional Von Neumann architecture.
Synthetic Biology. The science of altering genetics including by using gene manipulation and computer science techniques to create organism variations or artificial biological systems.

Transcendent. Going beyond ordinary limits or experience. To head for or reach a superior or supreme state.

Turing Test. A proposed test of the mathematician Alan Turing to demonstrate that a machine is intelligent. If a human cannot tell which responses from a human and from a machine to common inquiries and conversation are from the human or the machine, then the machine has passed the test and can be described as intelligent. This controversial test is widely used in artificial intelligence discussions.

Von Neumann architecture. The original and predominant digital computer architecture in which computers are constructed by separating memory and processing and operate by executing a series of equations or algorithms. Alternate computer architectures are sought where there is massively parallel real-time data inputs (such as from the eyes, ears, nose, and touch), where the algorithms are not fully known nor need to be fully understood, and the energy consumed by the computer must be kept to a minimum (see Synaptic Processor above).

Wiki. A collaborative website used as the basis for much collaborative intelligence and collective intelligence. The content on the wiki site can be added to or edited by anyone who has access to it much like an electronic community bulletin board. Wikis are usually focused on a particular topic such as pharmaceutical research and they can be public (accessible by anyone with access to the Internet) or private (limited to members of a company, agency or association).

REFERENCES

Chapter 1 New Creativity for New Prosperity

Bellis, Mary, "Liquid Paper—Bette Nesmith Graham (1922-1980)," *About.com:* Inventors, http://inventors.about.com/od/lstartinventions/a/liquid_paper.htm.

Silverman, Kenneth, *Lightning Man*, Alfred A. Knopf, New York, 2003. (Biography of Samuel F.B. Morse).

Vogelstein, Fred, "The Untold Story: How iPhone Blew Up the Wireless Industry," *Wired Magazine*, Issue 16.02, January 9, 2008.

Chapter 2 Learning from the Masters

Breese, James L., Jr., email to the author.

Bryan, Charles S., *Edison the Man and His Work*, Garden City Publishing Co., New York, 1926, pp 8-9, 45, 281.

Capra, Fritjof, *The Science of Leonardo*, Doubleday, New York, 2007, pp. 29-31, 34-35, 70-73,169.

Csikszentmihalyi, Mihaly, *Creativity: Flow and the Psychology of Discovery and Invention,* Harper Perennial, New York, 1997.

Dyer, F.L., and Martin, T.C.: *Edison: His Life and Inventions,* New York, 1910, Vol. I, p. 297; Vol. II, pp. 615-616.

Einstein, Albert, essay, *The World As I See It,* A. Einstein, *Ideas and Opinions, based on Mein Weltbild,* edited by Carl Seelig, New York: Bonzana Books, 1954 (pp. 8-11).

Fuller, Edmund, *Tinkers and Genius,* Hastings House, New York, 1955, p. 302.

Hughes, Thomas P., "How Did the Heroic Inventors Do It?" *American Heritage* (Magazine), Fall 1985, Vol. 1, Issue 2, p. 22.

Isaacson, Walter, *His Life and Universe* (about Einstein), Simon & Schuster, New York, 2007.

Kilham, Lawrence B. and Riley, David W., Apparatus and Method for Polymer Melt Stream Analysis, US Patent No. 4529306, July 16, 1985.

Music Encyclopedia On-Line.

New York Tribune, Editorial "The World," February 13, 1922.

Ohanian, Hans C., *Einstein's Mistakes,* Norton, New York, 2008.

Reti, Ladislao, Ed., *The Unknown Leonardo,* McGraw-Hill, New York, 1974, p. 300 (Folio Atlanticus 91v-a).

Runes, Dagobert D., ed., *The Diary and Observations of Thomas Alva Edison*, Philosophical Library, New York, 1948, 1976, p. 43.

Smolin, Lee, "Einstein's Lonely Path," *Discover Magazine*, September, 2004, p. 1, http://discovermagazine.com/2004/sep/einsteins-lonely-path.

Chapter 3 A Vision and Creative Process for Today's Inventors

Kilham, Lawrence B. and Riley, David W., Apparatus and Method for Polymer Melt Stream Analysis, US Patent No. 4529306, July 16, 1985.

Kilham, Lawrence B., *Great Idea to a Great Company: Making Inventions Pay*, Xlibris, 2001. Also available from Amazon.

Chapter 4 The Projected Mind: Your Key to Creativity

Chomsky, Noam, *Language and Mind*, Cambridge University Press, Cambridge, UK, 2006.

Chomsky, Noam, *Language and the Problems of Knowledge*, The MIT Press, Cambridge, Mass. 1988.

Csikszentmihalyi, Mihaly, *Creativity: Flow and the Psychology of Discovery and Invention*, Harper Perennial, New York, 1997.

Damasio, Antonio, *Descartes' Error*, Penguin, New York, 1994, pp. 106-107, 162-163.

Enard, Wolfgang et al, "Molecular evolution of FOXP2, a gene involved in speech and language," *Nature* 418, 22 August 2002, pp. 869-872.

Grandin, Temple, *Thinking in Pictures*, Vintage Books, New York, 1995, pp. 4, 11, 208, 210.

Hawkins, Jeff, *On Intelligence*, Holt, New York, 2004, pp. 181,183,177-205.

Jung, Rex E. et al, "Biochemical Support for the 'Threshold' Theory of Creativity: A Magnetic Resonance Spectroscopy Study," *Journal of Neuroscience*, 29(16), April 22, 2009, pp. 5319-5325.

Kandel, Eric R., *In Search of Memory*, Norton, New York, 2006, pp. 388-390.

Lorenz, Ralph and Sotin, Christophe, *Scientific American*, March, 2010, p. 36.

Minsky, Marvin, *The Emotion Machine*, Simon & Schuster, New York, 2006.

Schacter, Daniel L. et al, "Episodic Simulation of Future Events—Concepts, Data and Applications," *Annals of the New York Academy of Sciences*, vol. 1196, 2008, pp. 39-60.

Treffert, Darold A. and Wallace, Gregory L., "Islands of Genius," *Scientific American Mind* (special report), 2007.

Zaslow, Jeffrey, "What If Einstein Had Taken Ritalin? ADHD's Impact on Creativity." *The Wall Street Journal*, February 3,

2005. P. 6. For a detailed description of Ritalin, see http://en.wikipedia.org/wiki/Methylphenidate.

Chapter 5 The Amazing Brain: How it Works in the Creative Process

Carter, Rita, *The Human Brain Book*, DK Publishing, New York, 2009.

Crick, Francis H.C., "Thinking About the Brain," *Scientific American*, vol. 241, September 1979, pp. 181-188.

Damasio, Antonio, *Descartes' Error*, Penguin, New York, 1994, pp. 245-252.

Hesslow, G., "Conscious Thought as a Simulation of Behavior and Perception," *Trends in Cognitive Sciences*, vol. 6, pp. 242-247.

Kandel, Eric R., *In Search of Memory*, Norton, New York, 2006, pp. 159-161.

Kandel, Eric R., "Speaking of Memory," *Scientific American Mind*, October/November, 2008, pp.16-17.

Mitchell, Tom M. et al, "Predicting Human Brain Activity Associated with the Meanings of Nouns," *Science*, vol. 320, May 30, 2008, pp. 1191-1195. The text word corpus that Mitchell et al used is available on-line from a data bank at the University of Pennsylvania. It's "corpus" is composed of approximately one trillion word tokens analyzed alone, in pairs, threes, fours, and fives. The result is frequency of co-occurrences of all words. This data set was contributed by Google, Inc. based on analysis

of publicly accessible web pages. It is found at http://www.ldc.
upenn.edu/Catalog/CatalogEntry.jsp?catalogId=LDC2006T13.
Schacter, Daniel L. et al, "Episodic Simulation of Future
Events—Concepts, Data and Applications," *Annals of the New
York Academy of Sciences,* vol. 1196, 2008, pp. 39-60.

Squire, Larry R. and Kandel, Eric R., "Memory: From Mind to
Molecules," *Scientific American Library,* New York, 1999, p. ix.

Chapter 7 Artificial Intelligence: Can
It do our Inventing for Us?

Capra, Fritjof, *The Science of Leonardo,* Doubleday, New York,
2007, p. 169.

Conway, Flo and Siegelman, Jim, *Dark Hero of the Information
Age—In Search of Norbert Wiener the Father of Cybernetics,* Basic
Books, New York, 2005. (Good reading on the early days of AI,
cybernetics and computers.)

Gelernter, David, "Artificial Intelligence Is Lost in the Woods,"
Technology Review, July/August 2007, http://www.technologyreview.
com/article/408171/artificial-intelligence-is-lost-in-the-woods/.

IBM Corporation. "IBM Unveils Cognitive Computer Chips,"
August 18, 2011, http://www-03.ibm.com/press/us/en/pressre-
lease/35251.wss.
See also, Simonite, Tom, "Thinking in Silicon," MIT
Technology Review, January 2014, http://m.technologyreview.
com/featuredstory/522476/thinking-in-silicon/#.UtLxFQENdbw.
gmail.

Kurzweil, Ray. *The Singularity is Near: When Humans Transcend Biology,* Penguin Books, New York, 2005, pp. 71-72, 130-131, 260-273.

Lehrer, Jonah, "Out of the Blue," *Seed Magazine,* March 3, 2008. http://seedmagazine.com/content/article/out_of_the_blue/.

Markoff, John, "Brainlike Computers, Learning from Experience," *The New York Times,* December 28, 2013, http://www.nytimes.com/2013/12/29/science/brainlike-computers-learning-from-experience.html?pagewanted=all&_r=0.

Markoff, John, "IBM Develops a New Chip That Functions Like a Brain," *The New York Times,* August 7, 2014, http://www.nytimes.com/2014/08/08/science/new-computer-chip-is-designed-to-work-like-the-brain.html?emc=edit_th_20140808&nl=todaysheadlines&nlid=58190540&_r=0.

Los Alamos Laboratory history related to computers and super computers: http://www.lanl.gov/history/atomicbomb/computers.shtml.

Marcus, Gary F., *The Algebraic Mind—Integrating Connectionism and Cognitive Science,* MIT Press, Cambridge, MA, 2001. PP. 1-34.

Markram, Henry, "The Blue Brain Project," *Nature Reviews, Neuroscience,* 7, February 2006pp 153-160.
http://en.wikipedia.org/wiki/Blue_Brain_Project.

McCorduck, Pamela, *Machines Who Think,* A K Peters, Natick, Mass., Revised edition, 2004. (comprehensive overall review of artificial intelligence).

Merolla, P.A. et al. "A million spiking-neuron integrated circuit with a scalable communication network and interface," Science, Vol. 345 no. 6197 DOI: 10.1126/science.1254642, http://www.sciencemag.org/content/345/6197/668.full.

Minsky, Marvin, *The Emotion Machine,* Simon & Schuster Paperbacks, New York, 2007. (new perspectives on artificial intelligence).

Ramsden, Ed, "Just Add Parts and Shake...," *Sensors Magazine,* December 2006, pp. 32-33. http://www.sensorsmag.com/sensors-mag/just-add-parts-and-shake-1312.

"The Rise of Artificial Intelligence and Its Potential Effects on Business," (very good overview 8 minute video), http://www.technology-in-business.net/the-rise-of-artificial-intelligence-and-its-effects-on-business/.

"Watson (computer)," http://en.wikipedia.org/wiki/Watson.

Chapter 9 Finding Solutions in Oceans of Data

Benios, Thania, "At the Edge of Life's Code," *Scientific American,* April, 2008, pp. 106-109.

Cellucidate website: http://www.cellucidate.com/.

Harvard Medical School Fontana Laboratory website section on systems biology: http://fontana.med.harvard.edu/www/Documents/Lab/research.signaling.htm.

Kurzweil, Ray. *The Singularity is Near: When Humans Transcend Biology,* Penguin Books, New York, 2005, pp. 135-136.

Mullin, Rick, "Seeing the Forest at Pfizer," *Chemical & Engineering News*, American Chemical Society, September 3, 2007, p. 29.

Norvig, Peter, Director of Research, Google, Inc., http://norvig.com/spell-correct.html.

Whitefield, John, "Elucidate: Modeling by Community," *Santa Fe Institute Bulletin*, 2009, vol. 24, pp. 21-23.

Chapter 10 Organizing and Searching Data

Anderson, Chris, "The End of Theory: The Data Deluge Makes the Scientific Method Obsolete," *Wired Magazine*, June 23, 2008.

Baker, Stephen, "Google and the Wisdom of the Clouds," *Business Week*, December 13, 2007.

Berners-Lee, Tim, Hendler, James and Lassila, Ora, "The Semantic Web," *Scientific American*, May, 2001, pp. 90-97.

Crum, Chris, "Mayer of Google Talks Future of Search," *Webpronews.com*, September 11, 2008.

Hanson, David J., "Adapting to the Data Explosion," *Chemical & Engineering News*, pp. 25-29, August 10, 2009.

Norvig, Peter, "The Evolution of Web Search," *Technology Review*, January 2008, p. 32.

Shadbolt, Nigel and Berners-Lee, Tim, "Web Science Emerges," *Scientific American*, October 2008, pp. 76-81.

Stetter, Joseph R., Hayward, California, Private communication.

Chapter 11 Collective Intelligence: Managing
Large-Scale Research and Innovation

Bonabeau, Eric, "Decisions 2.0: The Power of Collective Intelligence," *MIT Sloan Management Review*, Winter 2009, pp. 45-52.

"The Climate Collaboratorium: A New Forum Brings Experts Together to Combat Global Warming," *MIT Sloan Management Review*, Spring 2009, pp. 20-21.

InnoCentive see http://www.innocentive.com/.

Malone, Thomas W. et al, "Harnessing Crowds: Mapping the Genome of Collective Intelligence," Working Paper No. 2009-001, MIT Center for Collective Intelligence, February 2009, http://cci.mit.edu/publications/CCIwp2009-01.pdf.

Malone, Thomas W., "How the MIT Climate CoLab Harnesses Collective Intelligence to Combat Climate Change," *MIT Sloan Management Review*, fall 2014, pp. 22-25.

MIT Center for Collective Intelligence, http://cci.mit.edu/. A good overview and library of working papers about collective intelligence.

Tapscott, Don and Williams, Anthony D., *Wikinomics—How Mass Collaboration Changes Everything*, Portfolio, New York, 2006, p. 226.

Chapter 12 The Future of MegaMinds

Greenfield, Susan, from "We are the Final Frontier," by Ian Sample, *The Guardian*, February 10, 2005, guardian.co.uk.

Hesse, Monica, "Truth: Can You Handle it? Better Yet: Do You know It When You See It," *The Washington Post*, April 27, 2008.

Weisenbaum, Joseph, *Computer Power and Human Reason*, Penguin Books, New York, 1984.

ACKNOWLEDGEMENTS

This book has required a lot of research and consultation to be as technically correct as possible as well as a pleasure to read. Fortunately here in Santa Fe, New Mexico there are a plethora of scientific experts and the following have been most generous with their time and thoughts: Chris Wood, The Santa Fe Institute; Bob Eisenstein, the Santa Fe Alliance for Science; and Stephen Guerin, The Santa Fe Complex, sadly now out of business. In Vermont, my friend John Schultz always had ready and fresh insights on the mathematical and philosophical issues. At MIT my professors emeriti Marvin Minsky and Murray Eden had a lot of value to add as did Professor Steve Eppinger. My professional colleague at KWJ Engineering, Joe Stetter, always was ready with excellent inputs as was my uncle, an imaginative engineer, Jim Breese. My neighbor Jeff Taylor, a patent and IP specialist, was very helpful in reviewing the second edition. I would like to thank my editor Tony Huston for all his inspired changes. Most important of all has been my wife Betsy who has put up with all this distraction and has had countless great suggestions.

ABOUT THE AUTHOR

Larry Kilham is from the third generation of a family which has produced notable inventors who built successful businesses. The author, a Sloan School of Management graduate from MIT, has three patents and has founded two high-tech companies. In 1986 Larry was the co-recipient of the IR-100 Award cited by *Research & Development* magazine for developing one of the 100 most significant technical products of the year. His experiences range from complex defense systems analysis to tiny device development.

Larry lives in Santa Fe, New Mexico and is keenly interested in AI, ecology, global resources and the science of complexity. More information about his books and his blog can be found at www.FutureBooks.info. He can be contacted at lkilham@gmail.com.